Accounting, Finance, Sustainability, Governance & Fraud: Theory and Application

Series Editor

Kıymet Tunca Çalıyurt, Iktisadi ve Idari Bilimler Fakultes, Trakya University Balkan Yerleskesi, Edirne, Türkiye

This Scopus indexed series acts as a forum for book publications on current research arising from debates about key topics that have emerged from global economic crises during the past several years. The importance of governance and the will to deal with corruption, fraud, and bad practice, are themes featured in volumes published in the series. These topics are not only of concern to businesses and their investors, but also to governments and supranational organizations, such as the United Nations and the European Union. Accounting, Finance, Sustainability, Governance & Fraud: Theory and Application takes on a distinctive perspective to explore crucial issues that currently have little or no coverage. Thus the series integrates both theoretical developments and practical experiences to feature themes that are topical, or are deemed to become topical within a short time. The series welcomes interdisciplinary research covering the topics of accounting, auditing, governance, and fraud.

Muharrem Kılıç · Sezer Bozkuş Kahyaoğlu
Editors

Algorithmic Discrimination and Ethical Perspective of Artificial Intelligence

Springer

Editors
Muharrem Kılıç
Human Rights and Equality Institution of Türkiye
Ankara, Türkiye

Sezer Bozkuş Kahyaoğlu
Commercial Accounting Department
University of Johannesburg
Johannesburg, South Africa

ISSN 2509-7873 ISSN 2509-7881 (electronic)
Accounting, Finance, Sustainability, Governance & Fraud: Theory and Application
ISBN 978-981-99-6329-4 ISBN 978-981-99-6327-0 (eBook)
https://doi.org/10.1007/978-981-99-6327-0

This Springer imprint is published by the registered company Springer Nature Singapore Pte Ltd.
The registered company address is: 152 Beach Road, #21-01/04 Gateway East, Singapore 189721, Singapore

Paper in this product is recyclable.

Acknowledgements

Algorithmic Discrimination and the Prohibition of Discrimination in the Age of Artificial Intelligence

Today's humanity is living in a new digital age where the globalization of digital technologies such as "artificial intelligence, internet of things, and robotics". Artificial intelligence (AI), which is thought to have emerged in the middle of the twentieth century at the same time as the historical past of modern computer technology, shows the reach of the new technical move of the age of intensive mechanization opened by the Industrial Revolution. Today, national governments, companies, researchers, and citizens live in a new "data world" where data is getting "bigger, faster and more detailed" than ever before. The AI-based "digital world order" which continues to develop speedily with technological advances, points to a great transformation from the business sector to the health sector, and educational services to the judicial sector. As a result of the development of AI as a creation of digital age technology, today humanity exists in an "*algorithmic society*" that imperviously surrounds individual, social life, and public space with all sectors. This modern algorithmic order we are in has transformative effects.

"Technological fetishism"; today, it is seen to take new forms such as "digital positivism", "big data fetishism", or "post-humanist ideology". In today's world where technological fetishism prevails, there are serious concerns about the protection of fundamental rights and freedoms and the balance of freedom and security. As a matter of fact, AI is directly linked to "health and safety, freedom, privacy, dignity, autonomy and non-discrimination", and these also include ethical concerns. Perhaps the most important risk of using AI systems in algorithmic decision-making is the potential to produce "biased" results.

All these developments make fundamental rights and freedoms more fragile in terms of human rights politics. In parallel with this situation, the significance of protecting fundamental rights and freedoms is increasing. For these reasons, it is significant that national human rights institutions carry out studies on AI and human rights. Although a number of different justice criteria have been developed in "algorithmic fairness" research in the last few years, concerns about AI technology are increasing. In this context, the issue of preventing discrimination arising from the use of AI has become a major research topic on the agenda of international human rights associations besides human rights institutions and other relevant institutions operating at the national and regional levels.

Hence, the main discussion on the use of AI-based applications focuses on the fact that these applications lead to algorithmic bias and discrimination. Within the context of the Human Rights and Equality Law No. 6701, 15 grounds of discrimination are listed, particularly the grounds of "gender, race, ethnicity, disability and wealth". It is seen that AI-based applications sometimes lead to discrimination based on gender; sometimes, based on race, religion, wealth, and health status. As an equality institution, it is crucial for national human rights institutions to combat algorithmic discrimination and develop strategies for it.

For this purpose, in cooperation with the Human Rights and Equality Institution of Türkiye (HREIT) and Hasan Kalyoncu University, the "International Symposium on the Effects of Artificial Intelligence in the Context of the Prohibition of Discrimination" was held on March 30, 2022, in Gaziantep. The symposium aims to raise awareness of human rights violations that the use of AI may cause within the scope of the prohibition of discrimination and understand the role of equality bodies in combating these violations.

This study, which is the output of this symposium, aims to draw attention to "bias and discrimination" in the use of artificial intelligence and deals with the subject in 13 chapters. I hope that the book *Algorithmic Discrimination and the Prohibition of Discrimination in the Age of Artificial Intelligence* covers AI technologies in a sophisticated and comprehensive way, from data protection to algorithmic discrimination, the use of AI in criminal proceedings to hate speech, predictive policing to meta-surveillance, will be useful. I would like to congratulate Dr. Kahyaoğlu and all contributing authors for their work and hope that the book will provide a much better understanding of algorithmic discrimination and its effects.

Prof. Muharrem Kiliç
Chairman of Human Rights
and Equality Institution of Türkiye,
Ankara, Türkiye

Contents

Editors and Contributors

About the Editors

Muharrem Kılıç After completing his law studies at Marmara University Faculty of Law, Kılıç was appointed as an Associate Professor in 2006 and as a professor in 2011. He has held multiple academic and administrative positions such as the Institution of Vocational Qualification Representative of the Sector Committee for Justice and Security, Dean of the Law School, Vice-Rector, Head of the Public Law Department, and Head of the Division of Philosophy and Sociology of Law Department. He has worked as a Professor, Lecturer, and Head of the Department in the Department of Philosophy and Sociology of Law at Ankara Yıldırım Beyazıt University Faculty of Law. His academic interests are "philosophy and sociology of law, comparative law theory, legal methodology, and human rights law". In line with his academic interest, he has scientific publications consisting of many books, articles, and translations, as well as papers presented at the national and international congresses. Among a selection of articles in Turkish, "The Institutionalization of Human Rights: National Human Rights Institutions"; "The Right to Reasoned Decisions: The Rationality of Judicial Decisions"; "Transhumanistic Representations of the Legal Mind and Onto-robotic Forms of Existence"; "The Political Economy of the Right to Food: The Right to Food in the Time of Pandemic"; "The Right to Education and Educational Policies in the Context of the Transformative Effect of Digital Education Technology in the Pandemic Period"; and "Socio-Politics of the Right to Housing: An Analysis in Terms of Social Rights Systematics" are included. His book selections include "Social Rights in the Time of the Pandemic: The Socio-Legal Dynamics of Social Rights" and "The Socio-Political Context of Legal Reason". Among the English article selections, "Ethico-Juridical Dimension of Artificial Intelligence Application in the Combat to COVID-19 Pandemics" and "Ethical-Juridical Inquiry Regarding the Effect of Artificial Intelligence Applications on Legal Profession and Legal Practices" are among the most recently published academic publications. He worked as a Project Expert in the "Project to Support the Implementation and Reporting of the Human Rights Action Plan", of which the Human Rights Department of the Ministry of Justice is the main beneficiary. He worked as a Project Expert in the "Technical Assistance Project for Increasing Ethical Awareness in Local Governments". He worked as a Project Researcher in the "Human Rights Institution Research for Determination of Awareness and Tendency project." He has been awarded a TUBITAK Postdoctoral Research Fellowship with his project on "Political Illusionism: An Onto-political Analysis of Modern Human Rights Discourse". He was appointed as the Chairman of the Human Rights and Equality Institution of Turkey with the Presidential appointment decision numbered 2021/349 published in the Official Gazette on 14 July 2021. He currently serves as the Chairman of the Human Rights and Equality Institution of Turkey.

Sezer Bozkuş Kahyaoğlu is an Associate Professor at the Finance Faculty of Economics and Administrative Science at Kyrgyz Turkish Manas University Bishkek, Kyrgyz Republic. He graduated from Bosporus University and earned the B.Sc. degree in Management. He had an MA degree in Money Banking and Finance from Sheffield University and a Certification in Retail Banking from Manchester Business School, both with a joint scholarship of the British Council and Turkish Bankers Association. After finishing her doctoral studies, she earned a Ph.D. degree in Econometrics from Dokuz Eylul University in 2015. She worked in the finance sector in various positions at the head office. She worked in KPMG Risk Consulting Services as a Senior Manager. Afterward, she joined Grant Thornton as a Founding Partner of advisory services and worked there in Business Risk Services. Afterward, she worked in SMM Technology and Risk Consulting as a partner responsible for ERP Risk Consulting. During this period, she was a Lecturer at Istanbul Bilgi University in the Accounting and Auditing Program and Ankara University Internal Control and Internal Audit Program. She worked as an Associate Professor at Izmir Bakircay University between October 2018 and March 2022. She has joined the University of Johannesburg, Department of Commercial Accounting, School of Accountancy in South Africa to do more international joint academic work. Her research interests mainly include Applied Econometrics, Time Series Analysis, Financial Markets and Instruments, AI, Blockchain, Energy Markets, Corporate Governance, Risk Management, Fraud Accounting, Auditing, Ethics, Coaching, Mentoring, and NLP. She has various refereed articles, books, and research project experiences in her professional field. She was among the 15 leading women selected from the business world within the scope of the "Leading Women of Izmir Project" which was sponsored by the World Bank and organized in cooperation with the Aegean Region Chamber of Industry (EBSO), Izmir Governor's Office, and Metropolitan Municipality.

Contributors

Ahmet Esad Berktaş Turkish-German University, Istanbul, Turkey;
The Ministry of Health, Ankara, Turkey

Abdülkadir Bilen Computer Engineering, Turkish National Police, Counter Terrorism Department, Ankara, Turkey

Benay Çaylak Istanbul University, Istanbul, Turkey

Selin Çetin Kumkumoğlu Istanbul Bar Association, IT Law Commission, Artificial Intelligence Working Group, Istanbul, Turkey

Olgun Değirmenci Faculty of Law, TOBB Economy and Technology University, Ankara, Turkey

Murat Volkan Dülger Faculty of Law, İstanbul Aydın University, Istanbul, Turkey

Saide Begüm Feyzioğlu The Ministry of Health, Ankara, Turkey;
Acettepe University, Ankara, Turkey

Merve Ayşegül Kulular İbrahim Chairman of HREIT-Human Rights and Equality Institution of Türkiye, Ankara, Turkey

Didem İncegil Business Administration at the Faculty of Economics and Administrative Sciences, Hacı Bayram Veli University, Ankara, Turkey

Sezer Bozkuş Kahyaoğlu Chairman of HREIT-Human Rights and Equality Institution of Türkiye, Ankara, Turkey;
Commercial Accounting Department, University of Johannesburg, Johannesburg, South Africa

İbrahim Halil Kayral Health Management Department, Izmir Bakircay University, Izmir, Turkey

Muharrem Kılıç Chairman of HREIT-Human Rights and Equality Institution of Türkiye, Ankara, Turkey;
Human Rights and Equality Institution of Türkiye, Ankara, Turkey

Cenk Konukpay Istanbul Bar Association Human Rights Center, Istanbul, Turkey

Ahmet Kemal Kumkumoğlu Istanbul Bar Association, IT Law Commission, Artificial Intelligence Working Group, Istanbul, Turkey

Lethiwe Nzama-Sithole College of Business & Economics, Department of Commercial Accounting, The University of Johannesburg, Johannesburg, South Africa

Ş. Barış Özçelik Faculty of Law, Bilkent University, Ankara, Turkey

Figen Çizmeci Şenel Faculty of Dentistry of Karadeniz Technical University, Trabzon, Turkey

Tamer Soysal EU Project Implementation Department, Ministry of Justice, Ankara, Turkey

Abbreviations

ADM	Automated Decision-Making
AEGIS	Incremental Science System Aggregation Autonomous Discovery
AHBS	Family Physician Information System
AI	Artificial Intelligence
AI HLEG	High-Level Expert Group
ALTAI	Assessment List for Trustworthy AI
Authority	Turkish Personal Data Protection Authority
CAHAI	Ad Hoc Committee on Artificial Intelligence
CAHAI	Council of Europe's Ad Hoc Committee on AI
CCD	Standardized Electronic Transactions
CEDAW	Convention on the Elimination of all forms of Discrimination Against Women
CNN	Convolutional Neural Network
CoE	Council of Europe
COMPAS	Correctional Offender Management Profiling for Alternative Sanctions
CPOE	Electronic Ordering System
CRC	Convention on the Rights of the Child
CRPD	Convention on the Rights of Persons with Disabilities
CT	Computed Tomography
CVFDT	Concept Adapting Very Fast Decision Tree
DDO	Digital Transformation Office
DT	Decision Tree
ECHR	European Convention on Human Rights
EDPB	European Data Protection Board
EDPS	European Data Protection Supervisory
EHRs	Electronic Health Records
eMAR	Electronic Drug Management Record
EMR	Patient Record
E-triage	Electronic Triage
EU	European Union

GDPR	European Union General Data Protection Regulation
GDPR	General Data Protection Regulation
HART	Harm Assessment Risk Tool
ICCPR	International Covenant on Civil and Political Rights
ICERD	International Convention on the Elimination of All Forms of Racial Discrimination
ICESCR	International Covenant on Economic, Social and Cultural Rights
ICO	Information Commissioner's Office
IoMT	Internet of Medical Things
IoT	Internet of Things
IPL-	Intelligence-led Policing
ISIS	Islamic State of Iraq and Syria
KDS	Decision Support System
KVKK	Law no. 6698 on Protection of Personal Data
LR	Logistic Regression
LSTM	Long Short-Term Memory
MHRS	Central Physician Appointment System
MLP	Multi-layer Perceptron
MNB	Multinominal Naive Bayes
MoH	Ministry of Health
NLP	Natural Language Processing
OECD	The Organisation for Economic Co-operation and Development
PDPL	Personal Data Protection Law
PredPol	The Predictive Policing Company
R&D	Research and Development
RF	Random Forest
SDG	Sustainable Development Goals
Seq2Seq	Sequence to Sequence
SGK	Social Security Institution
SMO	Sequential Minimal Optimization
SVM	Support Vector Machine
TİHEK	Human Rights and Equality Institution of Turkey
TOBB	University of Economics and Technology
TÜBİTAK	The Scientific and Technological Research Council of Türkiye
TÜSEB	Health Institutes of Türkiye
TUYZE	Turkish Health Data Research and Artificial Intelligence Applications Institute
UK	The United Kingdom
UN	The United Nations
UNCAT	Convention against Torture and Other Cruel, Inhuman or Degrading Treatment or Punishment
UNESCO	United Nations Educational, Scientific and Cultural Organization
USA	The United States of America

List of Figures

List of Tables

Part I
Introduction

Chapter 1
The Interaction of Artificial Intelligence and Legal Regulations: Social and Economic Perspectives

Muharrem Kılıç and Sezer Bozkuş Kahyaoğlu

Abstract With artificial intelligence, a facilitating change expectation emerges not only at work but also at home and in all social environments including digital. This situation opens the discussion of artificial intelligence applications and the legal interaction with an ethical perspective regarding humanity, which concerns every environment, including public and private areas. This study aims to reveal the situations that may arise if the algorithms that form the basis of artificial intelligence applications are biased and what needs to be done to prevent this in relation to the Fair, Accountable, and Transparent (FAT) approach. Based on the fact that artificial intelligence has a wide variety of dimensions, we try to contribute through its legal and economic perspectives.

Keywords Responsible artificial intelligence · Algorithmic discrimination · Ethics · Human rights · Algorithmic bias · Fair, accountable, and transparent (FAT) · Artificial intelligence systems

1.1 Introduction

Today, there is a significant increase in data as a result of the development of the data generation process on a global scale in every field and in every sector with high technology. When this situation is shaped by the competitive environment, the analysis of the said big data has become a basic need. Although there are many tools and techniques developed for this, they can all be defined as autonomous systems and Artificial Intelligence (AI) applications, even if they are at different maturity levels

---

M. Kılıç
Chairman of HREIT-Human Rights and Equality Institution of Türkiye, Ankara, Turkey
e-mail: muharrem.kilic@tihek.gov.tr

S. Bozkuş Kahyaoğlu (✉)
Commercial Accounting Department, University of Johannesburg, Johannesburg, South Africa
e-mail: sezer.kahyaoglu@uj.ac.az; sezer.bozkus@bakircay.edu.tr

M. Kılıç and S. Bozkuş Kahyaoğlu (eds.), *Algorithmic Discrimination and Ethical Perspective of Artificial Intelligence*, Accounting, Finance, Sustainability, Governance & Fraud: Theory and Application, https://doi.org/10.1007/978-981-99-6327-0_1

in general. While these innovations started the era of digitalization, they also brought new risks as well as opportunities. The value of big data is known all over the world and efforts are made for value-added creative approaches. However, this is not an easy value-creation process, on the contrary, it becomes a source of social concern due to increasing legal, social, and institutional risks (ARTICLE 19 2019). In this study, these risks, which tend to increase constantly, are discussed, and evaluations are made in the context of human rights, ethical principles, and sustainability issues in the social dimension, which is one of the key application areas of AI.

1.2 Algorithms and the Possibility of Algorithmic Bias

Algorithms have a key role in the functioning of processes based on AI applications (Danks and London 2017). Therefore, the possibility of algorithmic bias is emphasized, and concerns are expressed by revealing the risks in this regard. However, the discussions so far in this field appear within the framework of the different meanings and uses of the term "bias" and there is complete confusion. In the literature, it is sometimes used only as a descriptive term, and sometimes it is expressed as a negative term. These different and varied approaches can create confusion and hinder healthy debate about when and how to respond to algorithmic bias. It is a fact that there are real concerns about this situation that we should worry about. However, it is also important to note that many different topics are unhelpfully brought together under the heading of "algorithmic bias". In this respect, it seems that public debates about algorithmic bias combine many different types, sources, and impacts of these biases (Kirkpatrick 2016).

It is accepted that the word "bias", which we take as a starting point, has a negative connotation in the English language. Jackson (2021) defines an algorithm as "*a set of instructions or rules designed to perform a specific task,* duty, *or goal allowing computers, smartphones, websites, and the like to function and make desired decisions…*". In fact, bias emerges as a situation that should be avoided, and this situation is of course also problematic because it affects the decision-making quality (PwC 2021). On the other hand, we understand this term differently if we need to examine it more objectively: "bias" refers to the situation of deviation from a certain standard. This divergence can occur in a variety of ways, especially when describing statistical bias in which an estimate made in different domains deviates from a statistical standard. For example, "moral bias", such as deviation from true population value, or where a judgment deviates from a moral norm, and similarly for regulatory or "legal bias", "social bias", "psychological bias", and many others. From this point of view, if we make a generalization, we may encounter very different types of bias depending on the type of standard we used. The most important consequence to note here is that while the same thing may be viewed as biased by one standard, it may not be so by another (Danks and London 2017).

These different biases we mentioned for algorithms can originate from many different sources. Therefore, through this work, we try to contribute to society and

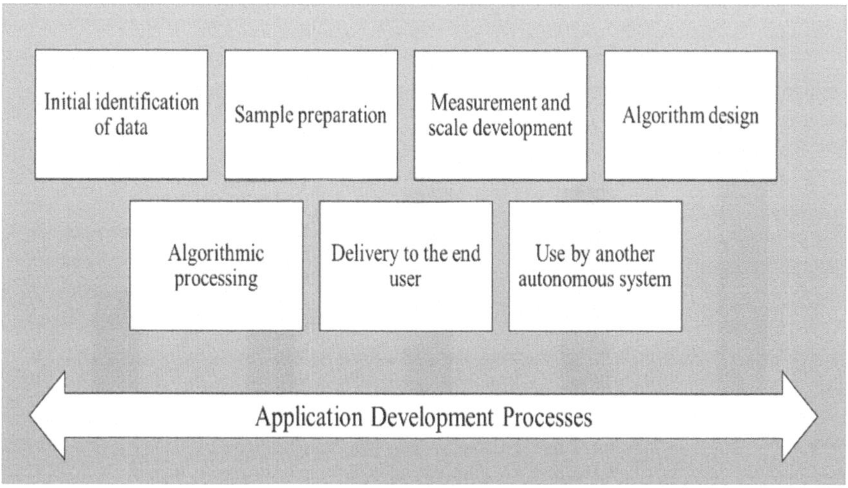

Fig. 1.1 The AI application development process open to algorithmic bias. *Source* Adapted from Kroll et al. (2016)

science and raise awareness about resolving different situations where an algorithm may become biased. Thus, we aim to provide a richer domain for assessing whether a particular bias deserves a response and, if so, what corrective or mitigation measures can be implemented. The areas where algorithmic bias is encountered in practice can be at every stage of the process as shown in Fig. 1.1 (Kroll et al. 2016).

This means awareness of how to respond to this issue in application development processes, from initial identification of data, sample preparation, measurement and scale development, algorithm design, algorithmic processing, delivery to the end user, or use by another autonomous system. Each independent entry point in the process of algorithmic bias provides the environment for analyzing a different set of problem-solving approaches, mindsets, and possibilities (Jackson 2021).

According to market research[1] conducted by PwC (2021), top managers think that measures should be taken to eliminate algorithmic bias. This idea is important in terms of developing responsible AI practices and highlighting the ethical dimension more. Top managers declared that this issue is among the top three among their corporate strategies in 2021 (Fig. 1.2).

As a summary of these considerations, detailed in Fig. 1.2, intelligent algorithms are needed for responsible AI (PwC 2021; Jackson 2021). This study aims to contribute to raising awareness of responsible AI applications. Especially the international community should build consensus and analyze in-depth how AI technologies violate human rights in different contexts. In particular, new approaches should be developed in order to result in effective legal solutions in emerging problem areas.

[1] Q: What steps, if any, will your company take in 2021 to develop and deploy AI systems that are responsible, that is, trustworthy, fair, bias-reduced and stable? Form list of 10 choices. Source: PwC (2021) AI Predictions. Base: 1,032.

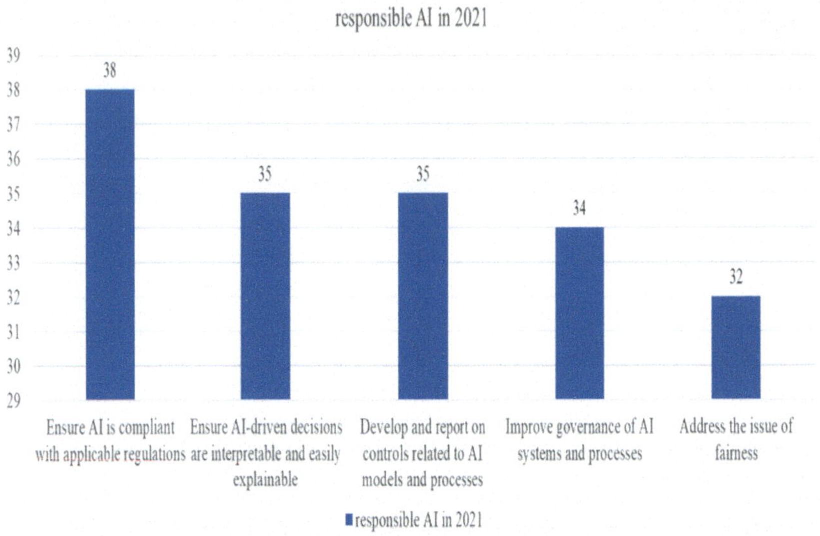

Fig. 1.2 Mitigating bias and creating responsible AI in 2021. *Source* PwC (2021)

It may be a mistake to limit the subject only to ethical principles (Saslow Kate and Lorenz 2017). Instead, it is important to focus on human rights by keeping our perspective wider than ethical principles. ARTICLE 19 (2019) argues that ethical principles remain weak within the scope of accountability and therefore will be insufficient to develop a responsible approach to AI applications. It emphasizes human rights as a solution to this. Particularly to have more accountability measures that can be taken and more liability can be provided for state and private sectors to implement responsible AI.

1.3 AI Systems with Fair, Accountable, and Transparent Perspective

The widespread use of AI systems in society day by day gives rise to concerns about discrimination, injustice, and exercise of rights, among others. There is a global scientific community that takes initiative in this regard and is working to address these issues by developing fair, accountable, and transparent (FAT) AI systems. This community aims to analyze many influencing factors, especially ethics, in order to expand an AI field of study based on the FAT perspective (ARTICLE 19 2019).

We prepared this book to deal with the key issues mentioned here in detail and to contribute to the relevant literature. The issue of prohibition of discrimination is discussed comprehensively with its legal and social dimensions. The concept of non-discrimination, which emerged in AI applications, is examined from a socio-political perspective. The rapid spread of AI applications as a result of digitalization also brings some social problems, especially in the judicial sector. Therefore, the "digital surveillance" or "meta-surveillance" mechanism and the responsible use of intelligent algorithms in the virtual environment are presented as a discussion platform for rethinking non-discrimination law in the age of AI (Borgesius 2018). In this context, the legal practices of different countries are evaluated by emphasizing the current situation and future expectations regarding the legal regulations regarding the prohibition of discrimination, which has become a strategic area in the age of AI. While we propose solutions to the problematic areas encountered in this field, we specifically state what needs to be done to design a successful contestation mechanism based on relevant literature. Accordingly, it is reminded that attention should be paid not only to the contestation itself or the structure of the algorithm alone but also to the underlying legal framework as well as the entire decision-making system, i.e., human, machine, and organizational mechanisms (Xiong et al. 2022). It should be noted that mitigating the violation of human rights, especially non-discrimination is important. In this respect, it is recommended to establish compliant mechanisms by comprising ex ante and ex post periods.

There is a need for regulating AI against discrimination. This issue should be discussed in a broad perspective starting from Data Protection Legislation to AI-Specific Measures. Depending on the concept of data protection, it generally accepts the right to protection of personal data as a fundamental human right. In this context, it imposes certain legal obligations on persons who have access to personal data to prevent the use of such data without the knowledge of the person concerned, and in some cases even without his or her consent. The most important risk emphasized here is the risk of being discriminated against during the processing of personal data by automated decision-making (ADM) systems. ADM systems are generally developed based on AI and machine learning technologies. Data of natural persons can be fed into the system to train the model used during the development of these applications. Therefore, personal data of natural persons is the basis for the decisions of ADM systems (Hofmann 2021).

Because of the rapid spread of AI applications, the new environment and problematic areas created by algorithms are evaluated. In particular, it is discussed whether the "right to disclosure" regulation in GDPR, which has significant effects in the legal sector and entered into force in the EU on April 25, 2018, can be used as a remedy to solve these problems. The problems of the social and legal environment in the digitalized should be assessed considering the new world order. In this context, it opens to discuss the level of sufficiency of fighting with the current criminal law rules regarding the use of AI from society's perspective. Based on the rapid spread of AI applications, it should be examined how responsibility can be handled in terms of criminal law in cases where the decisions taken constitute a crime.

First of all, it may come to mind that there is a criminal liability for the AI itself. However, some controversial issues should be noted. First, it is controversial that AI, which does not form legal personality, is held liable in terms of criminal law. Secondly, the responsibility of the software developer, which is important for AI applications and creating their algorithms. However, in this second case, there is a difference between the willful or negligent responsibility of the software developer, and this should be examined separately. In terms of the negligent responsibility of the software developer who created the AI algorithm, it is evaluated whether the issue of whether AI can be used in crime is predictable. Thus, in cases where an AI algorithm is used in the commission of a crime, it should be examined whether the existing regulations are sufficient in determining the responsibility in terms of criminal law.

Today, AI-oriented regulations are becoming more and more common in the field of law. In this context, predictive policing practices used in detecting and preventing possible crimes are gaining importance. In predictive policing practices, there are algorithms based on data about crimes committed in the past (such as the location and time of crimes, perpetrator, and victim) and a risk assessment and analysis regarding the commission of a new crime. The increase in the number of criminal data reveals a new trend in the prevention of crime and criminal activities before it occurs, with studies based on these data. When evaluated from this point of view, a security architecture emerges that emphasizes foresight within the framework of future studies by examining the data. In this respect, AI-oriented regulations are becoming widespread every day (Council of Europe 2019; Hofmann 2021).

As a result of this risk analysis, the police take the necessary measures to prevent crime. At first glance, the use of AI in predictive policing gives the impression that risk assessment and its consequences are independent of human bias. On the contrary, AI algorithms have the potential to contain all the biases of the people who designed the algorithms in question. In addition, the data analyzed based on algorithms are not free from prejudices and inequalities of societies. Therefore, predictive policing practices are likely to reflect the discriminatory thinking and practices of both individuals and societies. To prevent this situation, predictive policing practices are critically examined, and possible solutions are discussed in the further sections. In this study, the discriminatory side of predictive policing practices is revealed and, in this way, steps that can be taken to protect minorities and vulnerable groups in society are evaluated (McDaniel and Pease 2021).

Considering the criteria used in the software process with algorithmic bias in AI applications, it is highly likely to cause discrimination in legal aspects such as punishment, justice, and equality. Especially, in the process of using AI technologies by law enforcement, there are some basic problems that these discriminations can cause. These fundamental problems can be presented as cases that lead to discrimination and prejudice. In this framework, it is necessary to examine the fundamental issues surrounding fair trial rights related to transparency, accountability, and AI applications (Deloitte 2019). At this point, it can be stated that the need arising from the continuous updates that arise due to the progress and development of technologies should create a general framework in a way that covers the legal basis of the subjects in general and the time domain.

In particular, it is important for the sustainability of positive developments and improvements achieved through AI technologies. Because when used correctly and legally, AI technologies are extremely beneficial and efficient for both individuals and societies.

Discrimination can be expressed as the product of a decision-making mechanism using AI systems. Therefore, some legal problems may arise specific to such situations. Here is an example of these problems. It is a fact that the results produced by some AI technologies cannot be fully explained. However, in order to conclude that a decision is based on discrimination from a legal point of view, it is first necessary to know the reasons on which this decision is based. Considering the existing legal liability rules, those who develop and update the AI system cannot be held directly responsible to the victim of discrimination. It is necessary to produce a suitable solution to the problems in this regard. In this context, the general approach to be adopted by the legislators in the future related to the liability rules for the damages caused by AI will be decisive.

The interaction of law and AI is due to the fact that legal regulations and legal are the general rules that regulate every stage and stage of life. In this context, in every sector where AI is applied, AI legal interaction emerges depending on the legal regulations specific to that sector. This interaction reveals the necessity of addressing the application in each sector with its own scope and framework within the framework of AI law interaction. In general, it is seen that AI applications, which are evaluated within the framework of criminal law and public law, are subjects that need to be examined from a private law perspective, considering the sectors. In this context, discrimination, and the state of being a new product caused by the use of AI systems to this discrimination; emerges as a legal issue.

In practice, it is debatable whether the developers and updaters of AI systems have legal responsibility for the victims of discrimination. When evaluated from this point of view, it is necessary to produce a solution for this as well and to reveal information in decision-making.

In terms of the literature, it is seen that there are legal gaps regarding the damages caused by AI. More importantly, the information regarding the determination of responsibility and the punishment to be taken against it in the field of private law is still not fully revealed. In terms of country practices, the subject necessitates an important doctrinal discussion within the Anglo-Saxon and continental European legal systems. The emergence of technological advances mostly in countries with an Anglo-Saxon legal system causes these rules to be determined depending on the structure of the legal system in question. The process will initiate a transformation within the framework of the continental European legal system, as well as a new development depending on the harmonization efforts of the institutions that make up the legal rule (Atabekov 2023; Dixon 2022).

The most important field of change brought about by the development of AI is the discussion of the legal characteristics of robots that will provide the most important convenience in people's lives. In particular, the personality of the decision-making systems in matters that may affect people's lives, what the effects of those who develop these systems can be in the legal problems that arise, whether all possible factors are taken into account in the decision-making processes, reveal a wide range of information that can be considered as an important topic of discussion. From this point of view, the most important problem that robots and high technology will create if we call it automation in short, will have an impact on the employment structure of digital transformation in labor markets (Bavitz et al. 2019).

Today, it is seen that algorithm-based AI systems are used in the management of the process, including the recruitment process, to measure employee performance in human resources. Accordingly, it becomes controversial whether these systems produce an equal and fair result, especially ethical, by taking into account all corporate governance standards, elements, and criteria. In this sense, with AI technology, the necessity of considering the criteria and criteria of social rights emerges. At this point, the need for new jurisprudence to be formed in order to harmonize the previous court decisions with today's conditions is increasing. The interaction of AI and law, which is generally considered in the context of developments in the digital field, showed itself with the regulations aimed at preventing discrimination and hate speech on platforms such as Twitter, Instagram, Facebook, YouTube, and TikTok. The detection of this discrimination and hate speech is mainly based on AI methods, and techniques such as artificial neural networks, support vector machine algorithms, and decision trees are used in this framework. However, when the criteria of discrimination and hatred on the basis of discourse change over time, the adaptation and adaptation of these tools and the time difference that may arise between them may cause delays in implementation (Alarie et al. 2018). In this sense, it is necessary to take into account the effects of techniques aimed at reducing discrimination and hate speech, which may increase productivity by increasing the acceleration in other fields and applications, and also save time, together with measures to eliminate the aforementioned problems in digital environments (Cartolovni et al. 2022).

We can state that AI applications have the potential to initiate an important growth trend through positive internal and external economies, where economies of scale and economies of space are intertwined in economic terms. In this sense, it can be said that any AI application can create in another sector with another knowledge created by more than the size of the effect it has created in one sector.

In this context, the expectation of humanity about AI applications is formed according to the expectations, depending on the effects of these applications in the health sector and the quality of health services. Considering that climate change has become an important health problem in today's conditions, it can be stated that AI can initiate a great change in this field. It can be stated that collecting all information on the same platform, especially in health business processes, can reveal a new understanding of health through natural language processing methods and AI applications.

Especially, in the age of communication, the emergence of a global exchange of information in the field of medicine, while smart tools and robotic applications accelerate decision-making in diagnosis, it also enables many applications to be carried out remotely. The main problem here is the inadequacy of regulations against legal problems that may be caused by not fully determining the applications with algorithm-based criteria, despite the potential of AI applications to improve and reduce human error (Bavitz et al. 2019).

1.4 Concluding Remarks and Future Expectations

The legal dimension of AI applications in economic and social terms has a contrast and symmetry feature in itself. The contrast will be valid in practice with the rules and regulations based on the new algorithm and AI applications that will emerge as a result of teaching the rules and regulations to AI algorithms. In this case, the validity of the algorithmic bias and its outputs brings with it a discussion in itself.

The most important issue that we may encounter in terms of social and economic aspects, is whether the tools that learn, make decisions, and produce the decisions that have effects in the end, gain legal personality and the problem of legal penalties and also the sanctions against the negative results that arise here should be opened up for discussion.

The most important problem is that by teaching social and public knowledge to existing algorithms and AI systems, it becomes controversial who will have the power to determine how effective this will be in decision-making. Based on these principles, we come across the necessity of dealing with AI regulations with different dimensions from the current social development level and economic perspective. However, when evaluated with its self-feeding tendency, which creates positive social effects together with the spread effects, as mentioned before, a new architecture will emerge where scientists from various fields will come together to discuss future expectations.

References

Article 19 (2019) Governance with teeth: how human rights can strengthen FAT and ethics initiatives on artificial intelligence. https://www.article19.org/wp-content/uploads/2019/04/Governance-with-teeth_A19_April_2019.pdf. Accessed 09 May 2022

Alarie B, Niblett A, Yoon AH (2018) How artificial intelligence will affect the practice of law. Univ Tor Law J 68:106–24. https://tspace.library.utoronto.ca/bitstream/1807/88092/1/Alarie%20Artificial%20Intelligence.pdf. Accessed 01 Mar 2023

Atabekov A (2023) Artificial intelligence in contemporary societies: legal status and definition, implementation in public sector across various countries. Soc Sci 12:178. https://doi.org/10.3390/socsci12030178

Bavitz C, Holland A, Nishi A (2019) Ethics and governance of AI and robotics. https://cyber.harvard.edu/sites/default/files/2021-02/SIENNA%20US%20report_4-4_FINAL2.pdf. Accessed 1 Mar 2023

Borgesius FZ (2018) Discrimination, artificial intelligence, and algorithmic decision-making. Council of Europe-Directorate General of Democracy. https://rm.coe.int/discrimination-artificial-intelligence-and-algorithmic-decision-making/1680925d73. Accessed 09 May 2022

Cartolovni A, Tomicic A, Mosler EL (2022) Ethical, legal, and social considerations of AI-based medical decision-support tools: a scoping review. Int J Med Inform 161:104738

Council of Europe (2019) artificial intelligence and data protection. Adopted by the committee of the convention for the protection of individuals with regards to processing of personal data (Convention 108) on 25 January 2019. https://rm.coe.int/2018-lignes-directrices-sur-l-intelligence-artificielle-et-la-protecti/168098e1b7. Accessed 09 May 2022

Danks D, London AJ (2017) Algorithmic bias in autonomous systems. In: Paper presented at the proceedings of the 26th international joint conference on artificial intelligence. https://www.ijcai.org/proceedings/2017/654

Deloitte (2019) Transparency and responsibility in artificial intelligence. https://www2.deloitte.com/content/dam/Deloitte/nl/Documents/innovatie/deloitte-nl-innovation-bringing-transparency-and-ethics-into-ai.pdf. Accessed 09 May 2022

Dixon RBL (2022) Artificial intelligence governance: a comparative analysis of China, the European Union, and the United States. University of Minnesota Digital Conservancy. https://hdl.handle.net/11299/229505

Hofmann HCH (2021) An introduction to automated decision-making (ADM) and cyber-delegation in the scope of EU public law (June 23, 2021). University of Luxembourg Law Research Paper No. 2021-008, SSRN: https://ssrn.com/abstract=3876059 or https://doi.org/10.2139/ssrn.3876059. Accessed 10 May 2023

Jackson MC (2021) Artificial intelligence & algorithmic bias: the issues with technology reflecting history & humans. J Bus Tech L 16:299 (2021). https://digitalcommons.law.umaryland.edu/jbtl/vol16/iss2/5. Accessed 09 May 2022

Kirkpatrick K (2016) Battling algorithmic bias. Commun ACM 59(10):16–17

Kroll JA, Huey J, Barocas S, Barocas S, Felten EW, Reidenberg JR, Robinson DG, Robinson DG, Yu H, Accountable Algorithms (2016) University of Pennsylvania law review, vol 165, 2017 Forthcoming, Fordham Law Legal Studies Research Paper No. 2765268, SSRN: https://ssrn.com/abstract=2765268. Accessed 09 May 2022

McDaniel J, Pease K (2021). Predictive policing and artificial intelligence. Routledge Publications. ISBN 9780367701369

PwC (2021) Understanding algorithmic bias and how to build trust in AI. https://www.pwc.com/us/en/tech-effect/ai-analytics/algorithmic-bias-and-trust-in-ai.html. Accessed 09 May 2022

Saslow Kate K, Lorenz P (2017) Artificial intelligence needs human rights: how the focus on ethical AI fails to address privacy. Discrimination and Other Concerns

Xiong W, Fan H, Ma L, Wang C (2022) Challenges of human–machine collaboration in risky decision-making. Front Eng Manag 9(1):89–103 (2022). https://doi.org/10.1007/s42524-021-0182-0. Accessed 09 May 2022

Muharrem Kılıç After completing his law studies at Marmara University Faculty of Law, Kılıç was appointed as an associate professor in 2006 and as a professor in 2011. Kılıç has held multiple academic and administrative positions such as Institution of Vocational Qualification Representative of the Sector Committee for Justice and Security, dean of the law school, vice-rector, head of the public law department, head of the division of philosophy and sociology of law department. He has worked as a professor, lecturer, and head of the department in the Department of Philosophy and Sociology of Law at Ankara Yıldırım Beyazıt University Faculty of Law. His academic interests are "philosophy and sociology of law, comparative law theory, legal methodology, and human rights law". In line with his academic interest, he has scientific publications consisting of many books, articles, and translations, as well as papers presented in national and international congresses.

Among a selection of articles in Turkish, 'The Institutionalization of Human Rights: National Human Rights Institutions'; 'The Right to Reasoned Decisions: The Rationality of Judicial Decisions'; 'Transhumanistic Representations of the Legal Mind and Onto-robotic Forms of Existence'; 'The Political Economy of the Right to Food: The Right to Food in the Time of Pandemic'; 'The Right to Education and Educational Policies in the Context of the Transformative Effect of Digital Education Technology in the Pandemic Period' and 'Socio-Politics of the Right to Housing: An Analysis in Terms of Social Rights Systematics' are included. His book selections include 'Social Rights in the Time of the Pandemic: The Socio-Legal Dynamics of Social Rights' and 'The Socio-Political Context of Legal Reason'. Among the English article selections, 'Ethico-Juridical Dimension of Artificial Intelligence Application in the Combat to Covid-19 Pandemics' and 'Ethical-Juridical Inquiry Regarding the Effect of Artificial Intelligence Applications on Legal Profession and Legal Practices' are among the most recently published academic publications.

He worked as a Project Expert in the 'Project to Support the Implementation and Reporting of the Human Rights Action Plan', of which the Human Rights Department of the Ministry of Justice is the main beneficiary. He worked as a Project Expert in the 'Technical Assistance Project for Increasing Ethical Awareness in Local Governments'. He worked as a Project Researcher in the 'Human Rights Institution Research for Determination of Awareness and Tendency project.' Kılıç has been awarded a TUBITAK Postdoctoral Research Fellowship with his project on 'Political Illusionism: An Onto-political Analysis of Modern Human Rights Discourse.' He was appointed as the Chairman of the Human Rights and Equality Institution of Turkey with the Presidential appointment decision numbered 2021/349 published in the Official Gazette on 14 July 2021. He currently serves as the Chairman of the Human Rights and Equality Institution of Turkey.

Sezer Bozkuş Kahyaoğlu Assoc. Prof.Sezer Bozkuş Kahyaoğlu graduated from Bosporus University and earned the degree B.Sc. in Management. Sezer had MA degree on Money Banking and Finance at Sheffield University and Certification in Retail Banking from Manchester Business School, both with a joint scholarship of British Council and Turkish Bankers Association. After finishing her doctoral studies, Sezer earned Ph.D. degree in Econometrics from Dokuz Eylul University in 2015. Sezer worked in finance sector in various positions of head office. Sezer worked in KPMG Risk Consulting Services as Senior Manager. Afterwards Sezer joined to Grant Thornton as founding partner of advisory services and worked there in Business Risk Services. Afterwards Dr. Sezer worked in SMM Technology and Risk Consulting as Partner responsible from ERP Risk Consulting. During this period Sezer was a lecturer at Istanbul Bilgi University at Accounting and Auditing Program and Ankara University Internal Control and Internal Audit Program. Sezer worked as Associate Professor at Izmir Bakircay University between October 2018 and March 2022. Her research interests mainly include Applied Econometrics, Time Series Analysis, Financial Markets and Instruments, AI, Blockchain, Energy Markets, Corporate Governance, Risk Management, Fraud Accounting, Auditing, Ethics, Coaching, Mentoring and NLP. Sezer has various refereed articles, books, and research project experiences in her professional field. She was among the 15 leading women selected from the business world within the scope of the "Leading Women of Izmir Project" which was sponsored by the World Bank and organized in cooperation by Aegean Region Chamber of Industry (EBSO), Izmir Governor's Office, and Metropolitan Municipality.

Part II
Prohibition of Discrimination in the Age of Artificial Intelligence

Chapter 2
Socio-political Analysis of AI-Based Discrimination in the Meta-surveillance Universe

Muharrem Kılıç

Abstract The AI-based "digital world order" which continues to develop rapidly with technological advances, points to a great transformation from the business sector to the health sector, and educational services to the judicial sector. All these developments make fundamental rights and freedoms more fragile in terms of human rights politics. This platform of virtuality brings with it discussions of "digital surveillance" or "meta-surveillance" which we can define as a new type of surveillance. It should be stated that the surveillance ideology produced by this surveillance power technology also has an effect that we can describe as "panoptic discrimination". The use of these algorithms, especially in the judicial sector, brings about a transformation in terms of discrimination and equality law. For that reason, the main discussion on the use of AI-based applications focuses on the fact that algorithmic bias and discrimination. It is seen that AI-based applications sometimes lead to discrimination based on gender; sometimes based on race, religion, wealth, and health status. As an equality institution, it is significant for national human rights institutions to combating algorithmic discrimination and develop strategies for it.

Keywords Surveillance · Meta-surveillance · Artificial intelligence · Algorithmic discrimination · Human rights

2.1 Introduction

Today's humanity is living in a new digital age where the globalization of digital technologies such as the "*internet of things* (IoT), *artificial intelligence* (AI), *and robotics*". This digital age has significant effects on human rights. Our daily life, called techno-capitalism (Gülen and Arıtürk 2014: 117) is becoming technological day by day. With the development of digital technologies, AI which is effectively

M. Kılıç (✉)
Human Rights and Equality Institution of Türkiye, Ankara, Turkey
e-mail: muharrem.kilic@tihek.gov.tr

M. Kılıç and S. Bozkuş Kahyaoğlu (eds.), *Algorithmic Discrimination and Ethical Perspective of Artificial Intelligence*, Accounting, Finance, Sustainability, Governance & Fraud: Theory and Application, https://doi.org/10.1007/978-981-99-6327-0_2

used in all spheres of life has a threat to colonizing humanity. As Miguel Benasayag pointed out, this digital universe we are in is evolving toward "*algorithmic tyranny*". As a matter of fact, according to Benasayag, AI colonizes in many ways, from mass surveillance to predictive law enforcement to data-based social interactions (Benasayag 2021: 10).

The AI-based "digital world order" which continues to develop rapidly with technological advances, points to a great transformation from the business sector to the health sector, and educational services to the judicial sector. This virtuality platform which is embedded in the digital universe produces new life forms that can be conceptualized as "onto-robotic representation" by diversifying them at great speed (Kılıç 2021: 203). This platform of virtuality brings with it discussions of "digital surveillance" which we can define as a new type of surveillance. The impermeable structure of surveillance power which has divine references can be described as the "eye on earth" of the god in heaven. By creating the illusion that they are under constant surveillance, this "technological god" establishes an absolute surveillance power over society even more effective than the god in the sky (Foucault 2015: 252).

This realm of absolute power, which we can conceptualize as the "god of surveillance", is evolving toward a "*digital leviathan*". This divine power produced by the surveillance power leads to a situation that leaves no room for mistakes in human life. In other words, this divine power establishes an impermeable life mechanism that finds itself in famous Turkish poet Ismet Ozel's lines:

> You were the one who gave Adam the opportunity to make mistakes,
>
> I didn't know how much of his luck fell out,
>
> I was young and I was saying why there is no margin for mistakes in my life (Ozel 2012).

Thus, the surveillance power becomes absolute and produces an "ideology of bond" on individuals. It should be stated that the meta-surveillance ideology produced by this surveillance power technology also has an effect that we can describe as "panoptic discrimination". As a matter of fact, American political scientist Virginia Eubanks who says "*The future is here but no one can access it*" emphasizes that such technologies are the first used in the neighborhoods of the poor for surveillance purposes, and if they work, they will also be applied to the rich (Rossmann 2020).

Eubanks further states that "*Automated decision-making shatters the social safety net, criminalizes the poor, intensifies discrimination, and compromises our deepest national values*". According to her, automated decision-making reframes shared social decisions about who we are and who we want to be as systems engineering problems (Eubanks 2017: 12). In other words, it can be said that the technology of absolute surveillance power produces "class discrimination". These technological creations reproduce discrimination on the basis of class in a more categorical and impermeable way. Surveillance-based digital class production also makes the transition between traditional class impossible. In this respect, it should be pointed out that "surveillance-class discrimination" can be more dangerous.

All these developments make fundamental rights and freedoms more fragile in terms of human rights politics. Thus, it is seen that *digital colonialism* and *surveillance capitalism* produced by AI has reached levels that threaten human dignity (Mhlambi 2020: 3). In parallel with this situation, the significance of protecting fundamental rights and freedoms is increasing. For these reasons, it is crucial that national human rights institutions and equality bodies carry out studies on AI and human rights. Hence, the main discussion on the use of AI-based applications focuses on the fact that these applications lead to algorithmic bias and discrimination. Within the context of the Human Rights and Equality Law No. 6701, 15 grounds of discrimination are listed, particularly the grounds of "gender, race, ethnicity, disability and wealth". It is seen that AI-based applications sometimes lead to discrimination based on gender; sometimes based on race, religion, wealth, and health status. As an equality institution, it is crucial for national human rights institutions to fight algorithmic/ digital discrimination and develop strategies for it.

This paper will primarily focus on the surveillance phenomenon created by digital technologies that shape our lives gradually. Then, the use that AI especially in the judicial sector, will be mentioned. Usage of AI will be evaluated within the framework of the prohibition of discrimination. Also, the critical role played by human rights and equality institutions in the development of AI will be discussed. Finally, in addition to algorithmic discrimination, the phenomenon of robophobia will be mentioned.

2.2 A New Type of Surveillance: Meta-surveillance

George Orwell's (1903–1950) dystopian novel "1984" offers a narrative of the life of a totalitarian society trapped in a widespread surveillance network with the presence of "Big Brother" *who watches and records everything at any moment* (Orwell 2000). The phenomenon of surveillance was analyzed by Michel Foucault (1926–1984) as a power technique within the framework of its historical context. In Foucault's analysis of surveillance, the "panopticon" architecture of the English philosopher Jeremy Bentham (1748–1832) was the reference source. In his work titled *The Birth of Prison*, Foucault used the panopticon as a "metaphor of power" (Foucault 2011). Prison architecture, in which Jeremy Bentham believes detainees are under the gaze of an all-seeing inspector, is often thought of as a kind of prototype of enhanced electronic social incarceration (Lyon 2006: 23).

In the panoptic system, it is stated that the invisible surveillance structure of power creates mystification. Also, it is emphasized that this situation turns power into a fetish and helps the process of protecting the power itself, which is reflected in the social consciousness as an inaccessible, indivisible sacred structure for society (Çoban 2016: 119). According to the Director of the Surveillance Studies Research Center at the University of Kansas, William Staples; "*Postmodern surveillance is fragmentary, ambiguous, time–space-printed, and consumerist*". According to him, there are two main features of this surveillance. Firstly, this type of surveillance is based on an algorithmic assembly and control culture in everyday life. In this way, it is aimed to control people and punish them within the structure that Staples defines as "*meticulous rituals of power*". The second feature of postmodern surveillance is that it is a "preventive" mechanism for humanity (Staples 2014: 3).

On the other hand, panopticon is described as a "*crime-producing machine*". It is also expressed as a "*leviathan*", which further alienates people from themselves, oppresses and tries to destroy different thoughts (Çoban 2016: 126). Such that French philosopher Gilles Deleuze (1925–1975) uses strikingly the "*control society*" conceptualization for societies in which surveillance spreads like ivy rather than a tree rooting and growing in a relatively rigid, vertical plane like the panopticon (Deleuze 1992: 3).

As stated in the literature, it is seen with modernism that there is a transition from the panopticon, which expresses local surveillance based on coercion and pressure, to the synopticon and omnipticon, which define global "electronic surveillance" based on volunteerism and individual consent. The widespread use of social media tools and the internet also facilitates meta-surveillance all over the world. This is explained by concepts such as "*super-panopticon*" according to Mark Poster (1941–2012), "*liquid surveillance*" according to Zygmunt Bauman (1915–2017) and David Lyon (1948–), and "*digital siege*" according to Mark Andrejevic (Okmeydan Endim 2017: 59).

One of the transition processes of surveillance technologies is the phenomenon conceptualized by the sociologist Thomas Mathiesen (1933–2021) as the "Synopticon" created by modern television, in which many watch the few (Rosen 2004). In this context, the 2020 Brazilian science fiction series Omniscient is a striking example. The series presents a world vision in the city of São Paulo, where everyone is watched 24/7 by an individual drone that uses their AI and minimizes crime. This series is a unique example of the concept of omnipticon used by law professor Rosen (2004) in the sense of "*everyone spying on everyone, anytime and anywhere*".

It is emphasized that surveillance which David Garland calls "*the culture of control*" has become the symbol of many countries on a global scale, notably the United States of America (USA) and the United Kingdom (UK) (Lyon 2013: 28). Everyday surveillance, which David Lyon calls a disease peculiar to modern societies, is becoming increasingly widespread (Lyon 2013: 32). Parallel to all these development dynamics, especially China is regarded as the global driving force of "*authoritarian technology*". It is even claimed that Chinese companies are working directly with Chinese government officials to export authoritarian technology to like-minded governments to expand their influence and promote an alternative governance model. It is stated that China exports surveillance technology to liberal democracies as well as targets authoritarian markets (Feldstein 2019: 3).

Written by Virginia Eubanks who studies technology and social justice, the book *Automating Inequality* is discussing how government actors apply automated surveillance technologies that harm the poor and the working class. Virginia Eubanks states originally that high-tech tools have created a "*digital poorhouse*" that uniquely profiles the poor and the working class and deprives them of exercising certain rights (Eubanks 2017: 5).

It emphasized that some governments are currently using algorithmic systems to categorize people. In this context, the Social Credit System, which is used in

China provides an interesting example. This system collects data about citizens and scores them according to their social credibility. With this system, it is seen that some personal information of blacklisted people is intentionally made available to the public and is spotted online in cinemas and public places such as buses (Síthigh and Siems 2019: 5).

The message that the Chinese government wants to give through this system is quite clear: "*You are being followed algorithmically. Act accordingly*". In an interview conducted with a young Chinese girl within the context of the Coded Bias document, it is seen that this young girl stated that her social reliability score has a facilitating function in establishing personal relationships (Kantayya 2020).

It is said that the social credit score of the people in this system has an indicative role in human relations. Cathy O'Neil strikingly explains the perception of Chinese citizens toward this system as "*a kind of algorithmic obedience training*" (Kantayya 2020). This can be described as "*voluntary digital slavery*", inspired by Etienne de La Boétie's book (La Boetie 2020). Government use of AI tools to exploit efficiency gains can also be witnessed in China but on a completely different scale. China has been using machine learning powers to supercharge surveillance of the population and crack down on and control the Uyghur minority (Saslow Kate and Lorenz 2017: 10).

It is remarkably underlined such algorithmic systems lead to a situation that can be conceptualized as "socio-virtual exclusion". It can be mentioned the *Nosedive* episode of Black Mirror as a striking example (Wright 2016). Within this context, there is a world where people live in an environment where they can rate each other from 1 to 5 points; according to certain scores, friendship relationships are established and health care is provided, and these scores affect the socioeconomic status of people (Wright 2016). This series is an important example of such algorithmic structures leading to social exclusion. The use of technological devices based on AI causes a situation that distinguishes people, puts them in a certain class category, and leads to layered discrimination. In this context, surveillance discrimination is a phenomenon that is widely used on a global scale.

In conclusion, the "technological panopticon" era has begun, and today society is kept under control with different monitoring technologies and surveillance techniques (Yumiko 2008). So much so that surveillance, as Bauman points out, fluidly in a globalized world with, but exceeding, nation-states and emerges at a certain temporal and spatial distance (Bauman and Lyon 2016:16). As a result of the data revolution, it is seen that Big Brother has been replaced by Big Data. Thus, a "transparency society" emerges that records every detail of everyday life without a gap. As Byung Chul Khan described in his book "*Capitalism and the Impulse to Death*", this "society of transparency" creates a "digital panopticon" that produces a new surveillance technology (Han Chul 2021). Panopticism is one of the characteristic features of our modern society. It is a form of power applied to individuals in the form of transformation of individuals according to certain rules under the form of control, punishment, and reward within the framework of personal and continuous surveillance (Foucault 2011: 237).

2.3 Use of Artificial Intelligence in the Judicial Sector

The dizzying effect of AI-based technology, which exists in all areas of our daily life, affects the whole world and forces many sectors, including the legal sector to transform. AI as a superior technology is used in the judicial sector to improve judicial services and enable access to justice. It is envisaged that this "digital revolution" will continue to transform the legal sector until "AI-powered jurors to internet courts; AI robot lawyers to judges; and AI-powered features for contract or team management" (Kauffman and Soares 2020: 222, 223). Therefore, rights-based concerns about the use of AI judicial activities are increasing. In the face of these concerns, many new non-governmental organizations such as Algorithm Justice League have started to operate (The Law Society 2018: 11).

In this context, it is stated that criticisms of automated decision-making mechanisms are addressed in four main categories (The Law Society 2018: 11). First, these systems have the "potential for arbitrariness and discrimination". For instance, there is some evidence that the COMPAS (*Correctional Offender Management Profiling for Alternative Sanctions*) algorithm, which has been increasingly used by USA courts to estimate the likelihood of offenders committing crimes again since 2012, discriminates against African American defendants by using structural background data. This situation is also revealed in ProPublica's (*Non-Profit Newsroom, 2007*) *Machine Bias Report* published in 2016 (Angwin et al. 2016). This report found that black defendants were often predicted to be at a higher risk of recidivism than they actually were. Also, black defendants who did not recidivate over a 2-year period were nearly twice as likely to be misclassified as higher risk compared to their white counterparts (Larson et al. 2016).

Within the framework of these criticisms as an international non-governmental organization (NGO) Fair Trials recommends the prohibition of the use of predictive profiling and risk assessment AI systems in criminal justice. According to them, AI and automated systems in criminal justice are designed, created, and operated in a way that makes these systems susceptible to producing biased results. In addition, it is emphasized that profiling people with predictive policing practices and taking action without committing a crime violates the presumption of innocence, which is one of the basic principles of criminal proceedings. These profiles and decisions may include demographic information such as the actions of the people they are in contact with and even data about the neighborhood in which they live. This situation also constitutes a violation of the principle of individual criminal responsibility (Fair Trials, AI, algorithms & data).

Secondly, there are concerns about the "*legal accuracy*" of the decisions made by this system due to the complex nature of the algorithms. In addition, such AI-based judicial decision-making mechanisms have the potential to specialize in producing *bespoke judicial decisions* within the framework requested by the AI system creator. Third, algorithmic systems lack transparency in terms of "*justification of the judicial decisions*", which is a fundamental principle of justice (The Law Society 2018: 11).

In this regard, the Fair Trials's criticism that any system that has an impact on judicial decisions in terms of criminal justice should be open to public scrutiny is crucial. However, technological barriers and systems' deliberate efforts to conceal the mechanisms by which algorithms work for profit-oriented reasons make it difficult to understand how such decisions are made. At this point, the Case of Wisconsin should be pointed out. This case in which Eric Loomis, who was tried in the USA, was sentenced to 6 years in prison for being considered "high-risk" based on the COMPAS risk assessment, provides a substantial example. In this case, Loomis appealed by arguing that he should see the algorithm and get information about its validity to present arguments as part of his defense, but his request was rejected by the State Supreme Court (Woods 2022: 74).

As seen in the relevant case, people do not have the right to know why they are perceived as "high-risk". However, in judicial activities, it is necessary for the convict to understand the situation regarding his conviction and to share the relevant decision with the public by the principle of publicity. It should be underlined that the idea of non-visible algorithms that lead to the conviction of individuals is extremely problematic (Watney 2017).

In conclusion, judicial justice constitutes the most basic guarantee of democratic political societies. For this reason, it should be noted that the use of AI in the judicial sector is crucial. Analyzing the risks posed by AI for humanity emerges as an important issue, especially in the field of jurisdiction. As the use of AI in criminal justice systems continues to increase, it will tend to increase in these problematic areas. Therefore, it is substantial to establish rights-based AI systems.

2.4 Digital/Algorithmic Discrimination

The widespread use of AI-based technologies on a global scale has also led to discussions about "bias" in algorithmic decision-making. It is often discussed that the increasing use of AI technologies aggravates issues of discrimination. Also, the increasing capabilities and prevalence of AI and autonomous machines raise new ethical concerns (O'Neil 2017). In fact, American mathematician, data scientist, and author, Cathy O'Neil argues that "*there is no such thing as a morally neutral algorithm*" and that "*there is an ethical dilemma at the core of every algorithm*" (O'Neil 2017).

First of all, it should be pointed out that algorithmic discrimination is not only a technical phenomenon regulated by law but also a phenomenon that should be evaluated from a socio-cultural perspective. Thus, as Xavier Ferrer et al. have rightly emphasized, defining what constitutes discrimination is a matter of understanding specific social and historical conditions and the ideas that inform it, and needs to be re-evaluated according to its implementation context (Xavier Ferrer et al. 2021: 76).

In the context of social and historical conditions, three situations can be mentioned that lead to algorithmic discrimination. The first situation of algorithmic bias is through deviations in the training or input data provided to the algorithm. The input data that are used can be biased and thereby lead to biased responses for those tasks. In particular, it is stated that a "neutral learning" algorithm can yield a model that strongly deviates from real population statistics or a morally justified type of model. Indeed, in such a model the input or training data is biased in some way (Danks and London 2017: 2). A second situation of algorithmic bias is the differential use of information in the input or training data. A third situation of algorithmic bias is stated to occur when the algorithm itself is biased in various ways. In this context, it is stated that the most evident instance of algorithmic processing bias is the use of a statistically biased estimator in the algorithm (Danks and London 2017: 3).

Also, it can draw attention to the potential of algorithmic bias to have a more dangerous dimension than human bias. Since these technical mechanized systems do not carry an element of humanity, they pose more risks in terms of bias. At this point, Microsoft's chatbot on Twitter, Tay, offers a different example. It is seen that this boat is racist and misogynist over time with what it has learned from people. This is an example of algorithms reflecting a copy of the world (Leetaru 2016). Therefore, it should be emphasized that social development cannot be mentioned when it comes to technologies where ethics are ignored. In addition, the good or bad intentions of the mind that develops the algorithmic hardware of the system have the potential to turn into "*systematic bias violence*" (Kılıç 2021: 216).

In fact, algorithms decide everything in line with the instructions of those who code them. The system of dismissal of teachers in Houston can be mentioned as an example. It is seen that this system is biased in the dismissal of a teacher who received many awards in the past (Hung and Joy Liddicoat 2018). Similarly, algorithms that determine who will be interviewed, hired, or promoted in the field of employment produced biased results. To illustrate, it has been revealed that the algorithm developed by Amazon to view the CVs of the people it will employ in 2014 is systematically biased against women and reduces the CV grades of female candidates. Therefore, the algorithm was systematically removed after a short time since it produces gender bias (Goodman 2018).

Another example would be mentioned in the study by Princeton University researchers who used a word association technique using the AI system for over 2.2 million words in 2017. As a result of this analysis, it was revealed that European names were perceived more beautifully than African Americans. In addition, it was found that the words "woman" and "girl" were more likely to be associated with art rather than science and mathematics (Hadhazy 2017). It can be stated that this result reflects the social prejudice widely observed in the social sphere.

As an example of social bias, it can be mentioned an interesting event that occurred in Israel in 2017, which led to the arrest of a Palestinian man for writing "good morning" on his social media profile due to an error in the machine translation service. In the event, the AI-powered translation service translated the word as "damage them" in English or "attack them" in Hebrew. Then, the person was questioned by the Israel police officer. After the erroneous translation was noticed, the person was released (Berger 2017).

At this point, it would be mentioned in the Coded Bias documentary, in which the prejudices caused by algorithms, including Joy Buolamwini, the founder of the

Algorithmic Justice League, are dramatically addressed. As seen in the related documentary, working in front of a computer for a facial recognition study Boulamwini, after she realizes that the system knows very little about dark-skinned people, she is experimenting with a white mask, and this time she sees that the system's facial recognition rate is higher. So, she started to work in this field. It is seen that facial recognition technologies are banned in many states, especially in San Francisco and California, as a result of the speech she made in Congress, where she stated that the people who most benefit from these technologies are the white men who produce and release them (Kantayya 2020). Similarly, Big Brother Watch, pointing out that the police use facial recognition technology in the UK, states that the system used has neither legal basis nor supervision. Silkie Carlo, the director of Big Brother Watch, who stated that the police have established a new method and tried what will happen, emphasizes that "*No human rights experiments can be conducted*" (Kantayya 2020).

Last but not least, the article titled "*Algorithmic Discrimination Causes Less Moral Outrage than Human Discrimination*" written by Academician Yochanan E. Bigman; Desman Wilson; Mads N. Arnestad; Adam Waytz, and Kurt Gray in 2020 provides a significant example. In the relevant study, it was examined whether people are less outraged by algorithmic discrimination than human discrimination, and it was revealed that people are less morally outraged by algorithmic discrimination. This situation is conceptualized as *algorithm outrage asymmetry* (Bigman et al. 2018: 23).

As conclusion, discrimination led by algorithms is an inevitable reality. Algorithmic discrimination appears as a reflection of the socio-cultural dynamics of society. So much so that this virtual order created by the algorithmic mind is part of its mental processes. Therefore, it would not be appropriate to characterize algorithmic discrimination as a bias situation that arises solely from the machine.

2.5 The Role of Human Rights and Equality Institutions Combating Against Digital Discrimination

The idea of institutional guaranteeing of international human rights norms through protective mechanisms structured on a local-national scale tends to become widespread on a global scale. These mechanisms, which are structured as national human rights and equality institutions, act as a "bridge" between national practice and supranational dynamics within the framework of the mission of monitoring and supervising the national implementation of the founding human rights conventions to which it is a party. National human rights institutions are defined as independent and autonomous local structures established by public authorities and are liable for the protection and promotion of human rights (Kılıç 2022: 15–60).

As known, the main discussion on the use of AI-based applications focuses on the fact that algorithmic bias and discrimination. It is seen that AI-based applications sometimes lead to discrimination based on gender; sometimes based on race, religion, wealth, and health status. As an equality institution, it is substantial for national human rights institutions to fight algorithmic discrimination and develop transparency and related strategies for it.

In this context, as a first step toward ensuring transparency in the use of AI systems, it highlights the need for national authorities to achieve greater transparency in the use of AI systems by making a comprehensive and systematic map of the various ways in which AI systems are deployed in their regions. It is stated that the results of such mapping should be made publicly available and should constitute a first step toward ensuring enhanced transparency in the use of AI systems (Allen and Masters 2020: 21–23).

Also, national human rights institutions should conduct a legal "gap analysis" to understand how AI systems can be designed to protect and prevent human rights abuses, taking into account the principle of equality and non-discrimination. It is emphasized that equality bodies need to develop and facilitate inter-agency structures in collaboration with all other relevant regulatory institutional structures, as discriminatory AI systems affect various areas (Allen and Masters 2020: 21–23).

These institutional structures should ensure that they obtain technical expertise in AI, by involving computer scientists. They can be considered for organizing public awareness campaigns for organizations in the public and private sectors (Borgesius 2018: 30–31). Equality bodies could require public sector bodies to discuss with them any planned projects that involve AI decision-making about individuals or groups. For instance, an Equality Body could help to assess whether training data are biased. Also, these institutional structures require each public sector body to use AI decision-making about people to ensure that it has sufficient legal and technical expertise to assess and monitor risks. Also, mechanisms can be required to regularly assess whether their AI systems have discriminatory effects. Equality Bodies and human rights monitoring bodies can help to develop a specific method for a "*human rights and AI impact assessment*" (Borgesius 2018: 30–31). With the context of specific method, these institutional structures can also consider organizing conferences, round tables, or other events on discrimination risks of AI (Borgesius 2018: 30–31).

As a result, national human rights institutions, which have the mission of protecting and promoting human rights have a critical role in combating the discrimination created by AI. Studies can be carried out to develop national algorithmic anti-discrimination strategies, especially awareness-raising activities.

2.6 A New Type of Bias: Robophobia

Research on AI technology generally focuses on "algorithmic discrimination". Emphasizing the lack of literature in terms of the problem of *human misjudgment of machines*, Andrew Keane Woods, a professor at the University of Arizona, offers a perspective that will reverse the discussions on AI. Through this article, he argues that most of the legal research on AI focuses on how machines can be biased. While accepting the threat of algorithmic bias, Andrew specifically recommends the creation of strategies to combat the anti-robot perception in society. It focuses on "Robophobia" as a new type of bias. He states that there are diversified negative attitudes, judgments, and concerns about algorithms in society and that Robophobia emerges in different areas (Woods 2022: 54).

According to him, robophobia is a common condition in health care, litigation, military, and streets. It is stated that in healthcare, patients choose human diagnoses to computerized diagnoses, even when they are told that AI is more influential. Similarly, in litigation, lawyers are reluctant to rely on computer-generated results, even when they have been proven to be more accurate than human discovery results. While autonomous weapons promise to reduce the risk of serious human error in the military, there is said to be a legal movement to ban what are called "killer robots". It is stated that on the streets, robots are physically assaulted. According to Andrew, humans are biased against machines in various ways (Woods 2022: 56).

In this context, author David Gunkel seeks the answer to "*How to Survive the Robot Apocalypse*". He emphasizes that the ethical system is both machine-dependent and machine-independent. According to Gunkel, the "*machinic other*" is *typically portrayed in the form of an invading army of robots descending on us from the outer reaches of space sometime in the not-too-distant future* (Gunkel 2022: 70). He argues that it is the machine that constitutes the symptom of ethics "symptom" understood as that excluded "part that has no part" in the system of moral consideration. Also, he thinks that ethics which has been historically organized around a human or at least biological subject needs the machine to define the proper limits of the moral community even if it concurrently excludes such mechanisms from any serious claim on moral consideration (Gunkel 2022: 70).

As a result, the debate about algorithmic discrimination is evolving toward a new dimension. It is stated that not only machines cause biased results but also that people cause a new type of bias by behaving with bias against machines. It is emphasized that a state of fear lies at the root of people's prejudice against robotic creations.

2.7 Conclusion

The digital world order, shaped on the basis of AI, machine learning, and IoT leads to radical transformations from the education sector to the judicial sector, the health sector to the business sector. Digitalization caused by the data revolution stands before us as an inevitable reality. Therefore, the necessity of a revisionist human rights theory that will make the "digital renaissance" of the classical human rights doctrine possible, is obvious. The industrialization of AI-based surveillance (Mosley and Richard 2022) is creating a customized digital panopticon. Surveillance power has a function that overshadows (Jeffrey 2009) democratic politics and paralyzes the channels of social legitimacy.

Impermeable surveillance creates a climate of fear that we can conceptualize as "*panopticphobia*" which obscures the effective use of rights and freedoms. This surveillance *phobia* appears as social anxiety triggered by impermeable surveillance.

The essential distortion of *philia* (Aristoteles 2020) which constitutes the existential substance of man in the singular sense and leads to nihilism, evolves into *phobia*. *Phobia*, which deconstruction the inherent context and possibilities of sociability, creates an ever-deepening environment of anxiety and fear. The sense of social insecurity caused by this environment of anxiety and fear provokes the desire to spy on the "other". The constant vigilance created by the feeling of insecurity reinforces the "desire for surveillance".

It should be noted that all these surveillance technologies emerged as part of humanity's quest for objectivity. So much so that disbelief and distrust in the objectivity of human judgments bring along the search for a transhuman judgment. Thus, it is seen that mechanical judgment is superior to human judgment. Belief in objectivity also shows people's distrust of people. The uncanny and insecure attitude of man toward man points to "social rootlessness" in the existential sense. This feeling of insecurity, which has become stronger in terms of social psychology, has led to the search for new forms of existence in the meta-universe.

The artificial order created through the self-appointed authority of the mental world that created the algorithmic software gains immunity over the idea of the inherent objectivity of mechanization in terms of accountability and questionability. It can be characterized this situation as a new generation of "*algorithmic authoritarianism*". The danger of this situation is that, while there is a clear legitimacy debate/question regarding authoritarianism in the exercise of the entrusted powers, the new generation of algorithmic authoritarianism will become immune from it.

So much so that mechanization, which has no place for conscientiousness as a founding will that reveals what is human on the basis of reason and emotion, can have a devastating effect on the idea and practice of justice. However, it should be noted that there is a threshold of humanity that cannot be crossed here. The uniqueness of the productive creativity of homosapiens has been replaced by the monotypic mediocrity/artificiality of Robosapiens. A new era of "robotic colonization" is being witnessed. The age of digital colonization will perform a function that reduces the uniqueness of the human being, which is structured with the algorithmic mind of robotic devices, to singularity and standardization.

In conclusion, the use of these algorithms, especially in the judicial sector, brings about a transformation in terms of discrimination and equality law. At this point, the algorithms that reproduce the common memory, practice, or acceptance of the society should be designed on an ethical basis to prevent discrimination. In fact, algorithmic systems are a replica that reflects the basic acceptances and thoughts of societies. Therefore, due to the mathematical language and automatized technology of the algorithms, blind allegiance, or belief in them will produce a "discrimination dogmatism". It can be said that all these development dynamics have turned into "*algorithmic neo-positivism*" as a new phase of the positivist age based on the idea of objectivity. This algorithmic virtuality plane points to a new positivistic age.

The focus of discussions on the use of AI is that such technologies contain discriminatory attitudes. AI-based technologies sometimes have a discriminatory effect on the basis of gender, sometimes race, and religion. At this point, it should be emphasized that the work of equality and national human rights institutions in this area is important. As a matter of fact, these institutional structures, as anti-discrimination institutions, receive applications and carry out awareness-raising activities.

References

Allen R, Masters D (2020) Regulating for an equal AI: a new role for equality bodies. Equinet Publication, Brussels
Angwin J, Larson J, Mattu S, Kirchne L (2016) Machine bias. ProPublica
Aristoteles (2020) Nikomakhos'a Etik (trans Saffet Babür). Bilgesu Yayıncılık, Ankara
Bauman Z, Lyon D (2016) Akışkan Gözetim (trans Elçin Yılmaz). Ayrıntı Yayınları, İstanbul
Benasayag M (2021) The tyranny of algorithms: a conversation with Régis Meyran (trans Steven Rendall). Europa Editions, New York
Berger Y (2017) Israel arrests Palestinian because Facebook translated 'Good Morning' to 'Attack Them'. https://www.haaretz.com/israel-news/palestinian-arrested-over-mistranslated-good-morning-facebook-post-1.5459427. Accessed 29 Mar 2022
Bigman YE, Wilson D, Arnestad MN, Waytz A et al (2018) Algorithmic discrimination causes less moral outrage than human discrimination. Cognition, 81
Borgesius ZF (2018) Discrimination, artificial intelligence, and algorithmic decision-making. Council of Europe
Çoban B (2016) Gözün İktidarı. Üzerine Panoptikon: Gözün İktidarı içinde Su Yayınları, İstanbul
Danks D, London AJ (2017) Algorithmic bias in autonomous systems. In: Paper presented at the proceedings of the 26th international joint conference on artificial intelligence. https://www.ijcai.org/proceedings/2017/654
Deleuze G (1992) Postscript on the societies of control. The MIT Press, p 59
Eubanks V (2017) Automating inequality: how high-tech tools profile, police, and punish the poor. St. Martin's Press, New York
Fair Trials. AI, algorithms & data. https://www.fairtrials.org/campaigns/ai-algorithms-data/. Accessed 28 Mar 2022
Feldstein S (2019) The global expansion of AI surveillance. Carnegie Endowment for International Peace. https://carnegieendowment.org/2019/09/17/global-expansion-of-ai-surveillance-pub-79847
Foucault M (2011) Büyük Kapatılma (trans Işık Ergüden; Ferda Keskin). Ayrıntı Yayınları, İstanbul
Foucault M (2015) Hapishanenin Doğuşu (trans Mehmet Ali Kılıçbay). İmge Kitapevi Yayınları, Ankara
Goodman R (2018) Why Amazon's automated hiring tool discriminated against women. https://www.aclu.org/blog/womens-rights/womens-rights-workplace/why-amazons-automated-hiring-tool-discriminated-against
Gülen K, Aritürk MH (2014) Teknoloji çağında rasyonalite, deneyim ve bilgi: Sorunlar & eleştiriler. Kaygı Uludağ Üniversitesi Fen-Edebiyat Fakültesi Felsefe Dergisi 22(2):113–131
Gunkel DJ (2022). The symptom of ethics: rethinking ethics in the face of the machine. Hum-Mach Commun, 4
Hadhazy A (2017) Biased bots: artificial-intelligence systems echo human prejudices. https://www.princeton.edu/news/2017/04/18/biased-bots-artificial-intelligence-systems-echo-human-prejudices
Han Chul B (2021) Kapitalizm ve Ölüm Dürtüsü. (trans Çağlar Tanyeri). İnka Kitap, İstanbul
Hung KH, Joy Liddicoat J (2018) The future of workers' rights in the AI age. https://policyoptions.irpp.org/magazines/december-2018/future-workers-rights-ai-age/. Accessed 5 Mar 2022
Jeffrey GR (2009) Shadow government: how the secret global elite is using surveillance against you. WaterBrook Press, New York
Kantayya S (2020) Coded bias, USA

Kauffman ME, Soares MN (2020) AI in legal services: new trends in AI-enabled legal services. Serv-Oriented Comput Appl, 14

Kılıç M (2021) Ethical-juridical inquiry regarding the effect of artificial intelligence applications on legal profession and legal practices. John Marshall Law J 14(2)

Kılıç M (2022) İnsan Haklarının Kurumsallaşması: Ulusal İnsan Hakları Kurumları. Türkiye İnsan Hakları ve Eşitlik Kurumu Akademik Dergisi 5(8)

La Boetie E (2020) Gönüllü Kulluk Üzerine Söylev (trans Mehmet Ali Ağaoğulları). İmge Kitabevi, Ankara

Larson J, Mattu S, Kirchner L, Angwin J (2016) How we analyzed the COMPAS recidivism algorithm. https://www.propublica.org/article/how-we-analyzed-the-compas-recidivism-algorithm

Leetaru K (2016) How Twitter corrupted Microsoft's Tay: a crash course in the dangers of AI in the real world. https://www.forbes.com/sites/kalevleetaru/2016/03/24/how-twitter-corrupted-microsofts-tay-a-crash-course-in-the-dangers-of-ai-in-the-real-world/?sh=7e67f0e326d2

Lyon D (2013) Gözetim Çalışmaları: Genel Bir Bakış. (trans Ali Toprak). Kalkedon Yayıncılık, İstanbul

Lyon D (2006) Günlük Hayatı Kontrol Etmek: Gözetlenen Toplum (trans Gözde Soykan). Kalkedon Yayıncılık, İstanbul

Mhlambi S (2020) From rationality to relationality: Ubuntu as an ethical and human rights framework for artificial intelligence governance (Carr Center Discussion Paper Series No: 2020–009). https://carrcenter.hks.harvard.edu/files/cchr/files/ccdp_2020-009_sabelo_b.pdf. Accessed 5 Mar 2022

Mosley RT, Richard S (2022) The secret police: cops built a shadowy surveillance machine in Minnesota after George Floyd's murder. https://www.technologyreview.com/2022/03/03/1046676/police-surveillance-minnesota-george-floyd/. Accessed 15 Apr 2022

O'Neil C (2017) Why we need accountable algorithms. https://www.cato-unbound.org/2017/08/07/cathy-oneil/why-we-need-accountable-algorithms/

Okmeydan Endim S (2017) Postmodern Kültürde Gözetim Toplumunun Dönüşümü:' Panoptikon'dan 'Sinoptikon' ve 'Omniptikon'a. Online Acad J Inf Technol 8(30)

Orwell G (2000) 1984 (trans Celal Üster). Can Yayınları, İstanbul

Ozel I (2012) Bir Yusuf Masalı. Şule Yayınları, İstanbul

Rosen J (2004) The naked crowd: reclaiming security and freedom in an anxious age. http://www.antoniocasella.eu/nume/rosen_2004.pdf. Accessed 5 Mar 2022

Rossmann JS (2020) Public thinker: Virginia Eubanks on digital surveillance and people power. https://www.publicbooks.org/public-thinker-virginia-eubanks-on-digital-surveillance-and-people-power/. Accessed 5 Mar 2022

Saslow Kate K, Lorenz P (2017) Artificial intelligence needs human rights: how the focus on ethical AI fails to address privacy. Discrim Other Concernshttps://doi.org/10.2139/ssrn.3589473

Síthigh DM, Siems M (2019) The Chinese social credit system: a model for other countries? (EUI working paper law). https://cadmus.eui.eu/handle/1814/60424. Accessed 5 Mar 2022

Staples WG (2014) Everyday surveillance: vigilance and visibility in postmodern life. Rowman & Littlefield, United Kingdom

The Law Society (2018) Artificial intelligence and the legal profession. England and Wales, p 11

Watney C (2017) Fairy dust, Pandora's box… or a hammer. A J Debate, https://www.catounbound.org/2017/08/09/caleb-watney/fairy-dust-pandoras-box-or-hammer/. Accessed 5 Jun 2022)

Woods AK (2022) Robophobia. Univ Color Law Rev, 93

Wright J (2016) Black Mirror: 'Nosedive'

Xavier Ferrer X, Nuenen T, Such JM et al (2021) Bias and discrimination in AI: a cross-disciplinary perspective. IEEE Technol Soc Mag 40(2)

Yumiko I (2008) Technological panopticon and totalitarian imaginaries: the 'War on Terrorism' as a national myth in the age of real-time culture. https://www.nodo50.org/cubasigloXXI/congreso04/lida_180404.pdf. Accessed 15 May 2022

Muharrem Kılıç After completing his law studies at Marmara University Faculty of Law, Kılıç was appointed as an associate professor in 2006 and as a professor in 2011. Kılıç has held multiple academic and administrative positions such as Institution of Vocational Qualification Representative of the Sector Committee for Justice and Security, dean of the law school, vice-rector, head of the public law department, head of the division of philosophy and sociology of law department. He has worked as a professor, lecturer, and head of the department in the Department of Philosophy and Sociology of Law at Ankara Yıldırım Beyazıt University Faculty of Law. His academic interests are "philosophy and sociology of law, comparative law theory, legal methodology, and human rights law". In line with his academic interest, he has scientific publications consisting of many books, articles, and translations, as well as papers presented in national and international congresses.

Among a selection of articles in Turkish, 'The Institutionalization of Human Rights: National Human Rights Institutions'; 'The Right to Reasoned Decisions: The Rationality of Judicial Decisions'; 'Transhumanistic Representations of the Legal Mind and Onto-robotic Forms of Existence'; 'The Political Economy of the Right to Food: The Right to Food in the Time of Pandemic'; 'The Right to Education and Educational Policies in the Context of the Transformative Effect of Digital Education Technology in the Pandemic Period' and 'Socio-Politics of the Right to Housing: An Analysis in Terms of Social Rights Systematics' are included. His book selections include 'Social Rights in the Time of the Pandemic: The Socio-Legal Dynamics of Social Rights' and 'The Socio-Political Context of Legal Reason'. Among the English article selections, 'Ethico-Juridical Dimension of Artificial Intelligence Application in the Combat to Covid-19 Pandemics' and 'Ethical-Juridical Inquiry Regarding the Effect of Artificial Intelligence Applications on Legal Profession and Legal Practices' are among the most recently published academic publications.

He worked as a Project Expert in the 'Project to Support the Implementation and Reporting of the Human Rights Action Plan', of which the Human Rights Department of the Ministry of Justice is the main beneficiary. He worked as a Project Expert in the 'Technical Assistance Project for Increasing Ethical Awareness in Local Governments'. He worked as a Project Researcher in the 'Human Rights Institution Research for Determination of Awareness and Tendency project.' Kılıç has been awarded a TUBITAK Postdoctoral Research Fellowship with his project on 'Political Illusionism: An Onto-political Analysis of Modern Human Rights Discourse.' He was appointed as the Chairman of the Human Rights and Equality Institution of Turkey with the Presidential appointment decision numbered 2021/349 published in the Official Gazette on 14 July 2021. He currently serves as the Chairman of the Human Rights and Equality Institution of Turkey.

Chapter 3
Rethinking Non-discrimination Law in the Age of Artificial Intelligence

Selin Çetin Kumkumoğlu and Ahmet Kemal Kumkumoğlu

Abstract Irrespective of artificial intelligence ("AI") developments of our age, discrimination always found a way to influence our communities directly or indirectly. On the other hand, discrimination of individuals is prohibited as a reflection of the principle of equality, in the constitutions of contemporary societies and in international laws regulating the fundamental rights of the individual. Within the framework of the mechanisms improved depending upon this principle and constitutional basis, a legal struggle is waged against discrimination of individuals as a person or as part of a community. However, discrimination has reached a different dimension in the digital environment with the developing technologies, especially AI systems. Such systems bring new problems and the anti-discrimination rules and mechanisms generated for the physical world cannot completely find a solution. The discrimination of AI systems could arise from reasons such as biased data sets and AI models, insufficient training data, and human-induced prejudices. At the same time, due to the black box problem in AI systems, the fact that it is not always clear which inputs lead to which result, and which inputs are another problem in the detection of discriminatory results. On the other side, problems caused by different applications such as credit scoring, exam grade determination, face recognition, and predictive policing, in which individuals face explicitly discrimination due to AI systems, are also emerging more and more. However, individuals do not always know that they deal with an AI system, and they may or may not be subject to discrimination. For instance, it has been revealed that the algorithm used in visa applications in the UK ranks the applications as red, green, and yellow, yet classifies a certain group as persons with suspicious nationality, and applications made by persons from this nationality receive higher risk scores and are more likely to be rejected. Even though the use of AI systems in our socio-economic life has reached an indispensable point, the prevention of discriminatory results caused by these systems also triggers the

S. Çetin Kumkumoğlu (✉) · A. Kemal Kumkumoğlu
Istanbul Bar Association, IT Law Commission, Artificial Intelligence Working Group, Istanbul, Turkey
e-mail: selin.cetin@kecolegal.com

A. Kemal Kumkumoğlu
e-mail: kemal.kumkumoglu@kecolegal.com

M. Kılıç and S. Bozkuş Kahyaoğlu (eds.), *Algorithmic Discrimination and Ethical Perspective of Artificial Intelligence*, Accounting, Finance, Sustainability, Governance & Fraud: Theory and Application, https://doi.org/10.1007/978-981-99-6327-0_3

obligations of states in the context of fundamental rights. In this regard, the state has an obligation not to use discriminatory AI systems within the scope of its negative obligation; again, it has the obligation to ensure that private institutions or individuals cannot violate a fundamental right against other individuals in a way that is contrary to non-discrimination, within the scope of the positive obligation of the state. Furthermore, as in the process of protecting each fundamental right, the state will have an obligation to ensure that AI systems do not cause discrimination, and in case of discrimination, to eliminate the elements that cause it. In parallel with the development of AI technologies, it may be necessary to reinterpret existing rules and mechanisms or to establish new rules and mechanisms. In this sense, as in the draft Artificial Intelligence Act ("AIA"), having regard to the fact that AI has become a unique sector and its spheres of influence, organizing an institution specific to AI, and establishing an audit mechanism for developing and placing AI on the market, could be considered as an effective way to prevent and eliminate discrimination. In this regard, this Article aims to open a discussion about implementing newly emerging solutions to the Turkish legal framework, such as introducing a pre-audit mechanism as "AI—human rights impact assessment", establishing AI audit mechanisms, and notification to individuals that they are subject to discrimination.

Keywords Artificial intelligence · Non-discrimination · Human rights · Audit mechanisms

3.1 Introduction: Non-discrimination and Its Legal Grounds

While discrimination can occur either intentionally or unintentionally, by not being treated equally without a valid reason among people who are considered to be in equal status in a legal system, it is accepted that discrimination may also arise indirectly, by being treated equally without a valid reason among people who are considered to be in an unequal status (Karan 2015: 237–238). People have been subjected to discriminatory treatment in society from past to present, and this situation has revealed the need for certain protection mechanisms. Accordingly, it was met around the non-discrimination. Non-discrimination which bases on providing equal and fair opportunities to all people to access the opportunities available in society, ensures that people are not subjected to discriminatory treatment due to reasons such as race, color, religion, language, gender, age, political opinion, wealth, marital status, and disability (Karan 2015: 237). Non-discrimination eliminates the ambiguity that arises with the principle of equality regarding which subject and by which criteria equality will be achieved, and it is accepted as a fundamental principle that guarantees the principle of equality.

Although many international conventions deal with non-discrimination, they do not define it. It can be listed as follows: Article 14 of the European Convention on Human Rights (ECHR), which guarantees equal treatment in the exercise of the rights set forth in the Convention,[1] Protocol 12 to the ECHR, which guarantees equal treatment in the exercise of any right (including rights under national law),[2] International Covenant on Civil and Political Rights (ICCPR),[3] International Covenant on Economic, Social and Cultural Rights (ICESC),[4] International Convention on the Elimination of All Forms of Racial Discrimination (ICERD),[5] Convention on the Elimination of all forms of Discrimination Against Women (CEDAW),[6] Convention against Torture and Other Cruel, Inhuman or Degrading Treatment or Punishment (UNCAT)[7] and Convention on the Rights of the Child (CRC),[8] Convention on the Rights of Persons with Disabilities (CRPD)[9] and İstanbul Convention.[10]

The aforementioned international conventions do not specify in detail which actions constitute discrimination. According to Gül and Karan (Gül and Karan 2011: 25), non-discrimination, as regulated in international conventions, aims to protect a person and groups of persons against the use of an irrelevant and ineffective quality of a person or group of persons in terms of the treatment that they face as a determining factor in the treatment.

It is seen that the principles regarding non-discrimination are also included in the Turkish national legislation. Article 10 of the Constitution has adopted an approach based on the principle of equality, by the provision "*Everyone is equal before the law without distinction as to language, race, colour, sex, political opinion, philosophical*

[1] See Article 14 of ECHR "Prohibition of discrimination: The enjoyment of the rights and freedoms set forth in this Convention shall be secured without discrimination on any ground such as sex, race, color, language, religion, political or other opinion, national or social origin, association with a national minority, property, birth or other status." https://www.echr.coe.int/documents/convention_eng.pdf, Accessed Feb 28, 2022.

[2] See Protocol 12 to the ECHR, https://www.echr.coe.int/Documents/Library_Collection_P12_ETS177E_ENG.pdf, Accessed Feb 28, 2022.

[3] See International Covenant on Civil and Political Rights, https://www.ohchr.org/en/professionalinterest/pages/ccpr.aspx, Accessed 28, 2022.

[4] See International Covenant on Economic, Social and Cultural Rights https://www.ohchr.org/en/professionalinterest/pages/cescr.aspx, Accessed Feb 28, 2022.

[5] See International Convention on the Elimination of All Forms of Racial Discrimination, https://www.ohchr.org/en/professionalinterest/pages/cerd.aspx, Accessed Feb 28, 2022.

[6] See Convention on the Elimination of all forms of Discrimination Against Women, https://www.ohchr.org/en/professionalinterest/pages/cedaw.aspx, Accessed Feb 28, 2022.

[7] See Convention against Torture and Other Cruel, Inhuman or Degrading Treatment or Punishment, https://www.ohchr.org/en/professionalinterest/pages/cat.aspx, Accessed Feb 28, 2022.

[8] See Convention on the Rights of the Child, https://www.ohchr.org/en/professionalinterest/pages/crc.aspx, Accessed Feb 28, 2022.

[9] See Convention on the Rights of Persons with Disabilities, https://www.un.org/development/desa/disabilities/convention-on-the-rights-of-persons-with-disabilities/convention-on-the-rights-of-persons-with-disabilities-2.html, Accessed Feb 28, 2022.

[10] See İstanbul Convention, https://rm.coe.int/168008482e, Accessed Feb 28, 2022.

belief, religion and sect, or any such grounds".[11] According to Karan, Article 10 of the Constitute is not specific to a particular type of non-discrimination, it can provide protection to different forms of discrimination via a broad interpretation (Karan 2015: 244).

As another national regulation, Law No. 6701 on the Human Rights and Equality Institution of Turkey ("TİHEK"), the principle of equality and the general aspects of non-discrimination have been determined in Article 3, types of discrimination have been listed in Article 4 and the scope of the prohibition of discrimination has been dealt with in Article 5. Additionally, provisions on non-discrimination are included in various regulations. For instance, pursuant to Article 5 titled the principle of equal treatment of the Labor Law No. 4857, no discrimination based on language, race, sex, political opinion, philosophical belief, religion, and sex or similar reasons is permissible in the employment relationship. Similarly, in accordance with Article 3 titled the principle of equality treatment before the law of the Turkish Penal Law No. 5237, in the implementation of the Penal Law, no one shall receive any privilege and there shall be no discrimination against any individual on the basis of their race, language, religion, sect, nationality, color, gender, political (or other) ideas and thought, philosophical beliefs, ethnic and social background, birth, and economic and other social positions. Pursuant to Article 82 titled Prohibition of Regionalism and Racism of the Political Parties Law No. 2820, political parties cannot pursue the aim of regionalism or racism in the country, which is an indivisible whole, and cannot engage in activities for this purpose. Pursuant to subparagraph d of paragraph 1 of article 4 titled General Principles of Social Services and Child Protection Agency Law No. 2828, class, race, language, religion, sect, or region discrimination cannot be paid regard to the execution and delivery of social services, in case the service demand is higher than the service supply, the priorities are determined on the basis of the degree of need and the order of application or determination.

Overall, non-discrimination law has legal grounds such as constitution, international contracts, and local laws in Turkey. However, with newly emerging AI systems, existing legal grounds and their applications seem to fall short in the context of combatting discrimination.

3.2 New Dimensions of Non-discrimination Emerge with AI

Developing technologies not only provide various conveniences and benefits for individuals and society but also differentiate the emergence of discrimination in society and its reflections on individuals. Undoubtedly, human decision-making includes mistakes and biases. However, when the same biases are presented by AI decision-making, it could have a much larger effect and discrimination for many people

[11] The principle of non-discrimination and equality have been generally accepted as positive and negative aspects of the same principle. See the details Karan (2015), p. 237; Ramcharan (1981), p. 252.

(European Commission 2020a, b: 11). This situation requires to deal with non-discrimination with new dimensions. To understand the fundamentals of discrimination that occurs through AI systems, an examination starting from the development stage of these systems will make the subject more comprehensive and understandable.

AI systems can technically cause discrimination in several ways.[12] For instance (Council of Europe 2018: 17), supposing that a company chooses "rarely being late " as a class label (Kartal 2021: 360–361) to assess if an employee is "good". However, if it is not considered that on average, low-income groups rarely live in the city center and must travel further to their work than other employees, and they are late for work more often than others due to traffic problems, that choice of class label would cause disadvantage to poorer people, even if they outperform other employees in other aspects.

On the other hand, the training data[13] can reflect discriminatory human biases. These can reproduce those same biases in AI systems. In an event that happened in the UK in the 1980s (Barocas and Selbst 2016: 682; Council of Europe 2018: 18), a medical school received too many applications, and as a solution, the school developed a computer program which has training data based on the admission files from earlier years. However, these training data were biased against women and people with an immigrant background. Therefore, the computer program discriminated new applicants based on those same biases.

Some cases result in indirect discrimination. It is mostly difficult to catch on them for each case because this type of discrimination mostly remains hidden. For example (Council of Europe 2018: 36), a person applied for a loan on the website of a bank that uses an AI for such requests. When the bank rejects a loan on its website, the person does not see the reason why it was rejected. Even if the person knows that an AI system is decided on the merits of his/her application, it would be difficult to discover if the AI system is discriminatory.

In addition, the discrimination problem can arise from the quality of data[14] to build AI systems. They may not represent the population it is used for. This can occur due to certain problematic data sources relating quality of data (European Union Agency for Fundamental Rights 2019: 14). The Guidance published by the UK government has noted that it can be assessed if a dataset has high enough quality with a combination of accuracy, completeness, uniqueness, timeliness, validity, sufficiency, relevancy, representativeness, and consistency (UK Government 2019). In this context, to evaluate the quality of data, some questions could be the guideway. For example (European Union Agency for Fundamental Rights 2019: 15), what information is

---

[12] According to Barocas and Selbst, these ways can be listed as follow: (i) how the "target variable" and the "class labels" are defined; (ii) labeling the training data; (iii) collecting the training data; (iv) feature selection; and (v) proxies. See details Barocas et al. (2016); CAHAI (2020a).

[13] Training data is defined as "*data used for training an AI system through fitting its learnable parameters, including the weights of a neural network*", in Article 3 of Regulation laying down harmonized rules on artificial intelligence (Artificial Intelligence Act).

[14] The notion of data quality refers to three aspects: (1) the characteristics of the statistical product, (2) the perception of the statistical product by the user, and (3) some characteristics of the statistical production process. See details: European Commission, Eurostat (2007).

included in the data? Is the information included in the data appropriate for the purpose of the algorithm? Who is covered in the data? Who is under-represented in the data?

On the other hand, the fact that the results of AI cannot be understood due to the technically complex structure is another risk. This situation, which is expressed as this black box situation was explained by Yavar Bathaee as follows (Bathaee 2018: 893):

> If an AI program is a black box, it will make predictions and decisions as humans do, but without being able to communicate its reasons for doing so. (...) This also means that little can be inferred about the intent or conduct of the humans that created or deployed the AI, since even they may not be able to foresee what solutions, the AI will reach or what decisions it will make.

Therefore, the unexplainable results of AI systems may complicate the understanding of their effects and—if any—the prevention of negatives.

Consequently, it can be stated that laws and methods we used to utilize for the prevention of discrimination in the physical world will be insufficient in the era of AI.

3.3 The Need for Explainable AI Systems

Designing, implementing, and deploying an AI system that respects principles of explainability, is critical both to detect the existence of discrimination and the reasons of it, its victims, and the responsible parties, in order to combat discrimination.

As mentioned above, discrimination can occur based on different technical reasons, and it can be possible to see its presence in every field. It may lead to consequences that hinder the exercise of fundamental rights—even if we don't realize them. For instance, the research in the Journal of the American Medical Informatics Association (Federal Trade Commission 2021) has noted an AI prediction model for the COVID-19 pandemic. In this context, in case the models use data that reflect existing racial bias in healthcare delivery, AI may cause healthcare disparities for people of color.

To find out the sources of the discriminatory problem, and more importantly, to mitigate and prevent the risk of emergence of these discriminatory problems of AI, there is a need for explainable AI systems.

Different communities such as philosophy, law, and political science can diversely deal with technical meanings of explainability. For example, in the report of the US National Institute of Standards and Technology, they introduce four principles for explainable AI: explanation, meaningful, explanation accuracy, and knowledge limits (National Institute of Standards and Technology 2021: 2–3). It is highlighted that explainability is a critical property for AI systems, to support system trustworthiness (National Institute of Standards and Technology 2021: 1).

An exact definition on the explainability of AI is not yet consolidated (Mecham et al. 2019: 2; Gohel et al. 2021: 2; Saeed and Omlin 2021: 1). One of the definitions is as.

> set of processes and methods that allows human users to comprehend and trust the results and output created by machine learning algorithms (Turri 2022).

Explainability is a preferred solution to mitigate a system's lack of transparency (Chazette et al. 2021: 1) and provide a tool to answer critical how and why questions about AI systems (Turri 2022). In this context, researchers indicate that to achieve trustworthiness of a machine, detailed "explanations" of AI decisions seem necessary (Doran et al. 2017: 1).

As mentioned by Felzmann et al. (2020), transparency-as-information[15] in AI is linked to explanations and explainability, and a person should be able to have access to prior information about the quality or intent of a decision-making process, and ex post about the outcomes and how they came about (Felzmann et al. 2020: 3337). Therefore, explainability is a requirement for designers and developers in order to enhance system robustness and to prevent bias, unfairness, and discrimination. It is also related to issues of user rights and technology acceptance (Confalonieri et al. 2020: 2).

In other respects, the legal community has approached explanation and explainability in recent years, from the perspective of the General Data Protection Regulation; for instance, the right not to be subject to a decision based solely on automated processing (Article 22) or the right to obtain from the controller confirmation as to whether or not personal data concerning him or her are being processed, and, where that is the case, access to the personal data (Article 15) (Edward and Veale 2017: 18–84; Casey et al. 2019; Sovrano et al. 2021). However, it is obvious that AI systems have their own specific dynamics, risks, and effects, and need unique approaches in the meaning of legislation. Therefore, it should be also evaluated in its own nature, in the context of explainability.

Legal scholars have debated whether and what kind of explanation the General Data Protection Regulation's (GDPR) right not to be subject to certain kinds of fully automated decisions (Article 22) requires, in light of the right of individuals to obtain "meaningful information about the logic involved, as well as the significance and the envisaged consequences" (Article 15(1)(h) GDPR) of the automated processing occurring.

Due to its social and individual importance and effects on explainability, transparency, and discrimination issues, numerous guidelines were published by various local, regional, and international institutions in recent years, for ensuring the key requirements concerning the development, deployment, and use of AI systems. In 2019, High-Level Expert Group ("AI HLEG") presented the term "Trustworthy AI"[16]

---

[15] According to Heike Felzmann et al., "*In much of the literature on transparency, the emphasis of transparency as a positive force relies on the informational perspective which connects transparency to the disclosure of information*". *See* details: Heike Felzmann et al., p. 3337.

[16] AI HLEG has noted that "*Trustworthy AI has three components, which should be met throughout the system's entire life cycle: (1) it should be lawful, complying with all applicable laws and*

as a new concept (European Commission 2019: 12). AI HLEG listed several principles of Trustworthy AI as (i) *Respect for human autonomy*, (ii) *Prevention of harm*, (iii) *Fairness*, and (iv) *Explicability*. In addition, to execute these principles, the AI HLEG has determined seven requirements: (1) *human agency and oversight*, (2) *technical robustness and safety*, (3) *privacy and data governance*, (4) *transparency*, (5) *diversity, non-discrimination, and fairness*, (6) *environmental and societal well-being*, and (7) *accountability* (European Commission 2019: 14). As is seen, "transparency" and "diversity, non-discrimination and fairness" are positioned as a guiding principle.

The following year, in 2020, AI HLEG focused on a practical tool for the AI systems under development. In this context, the AI HLEG presented the Assessment List for Trustworthy AI (ALTAI) for self-evaluation purposes, in order to assist organizations, understand what risks an AI system might generate, and how to minimize those risks while maximizing the benefit of AI (European Commission 2020a, b: 3).

In addition to AI HLEG documentation, other organizations and governments have also published their own principles in previous years (OECD 2019; Australian Government, Australia's AI Ethics Principles; Integrated Innovation Strategy Promotion Council of Japan 2019; Government of Canada 2021). All these instruments mainly point out the discrimination risks during the development, the deployment, the usage of AI, and adopt *non-discrimination* as one of the cornerstone principles.

3.4 The Responsibilities of the State and the Need for a New Legislative Framework

Discrimination is a phenomenon that can result from an act of a state actor, of a real person, or of a legal person. It is also valid in the concept of AI applications. First, a state has a negative obligation of not causing a discrimination; an obligation "not to do". A state itself has an obligation to avoid violations, which requires that it does not cause human rights violations (Gül and Karan 2011: 140). In an example where this obligation can be discussed, in 2020, the UK's Office of Qualifications and Examinations Regulation relied primarily on two pieces of information to calculate A-level[17] grades of students by using the algorithms: the ranking of students within a school and their school's historical performance. However, depending upon several factors, the algorithm placed so much importance on historical performance of a school, and it caused problems for high-performing students at underperforming

regulations, (2) it should be ethical, ensuring adherence to ethical principles and values, and (3) it should be robust, both from a technical and social perspective since, even with good intentions, AI systems can cause unintentional harm". See details: European Commission (2019).

[17] An "advanced level" or "A-level" is the final exams taken before university, and universities make offers based on students' predicted A-level grades. For details see Nidirect, As and A levels, https://www.nidirect.gov.uk/articles/and-levels#toc-0, Accessed Feb. 18, 2022.

schools. As a result of the algorithmic bias, most students have gotten lower grades even from their teachers' estimations (Porter 2020).

Second, a state has a positive obligation, an obligation "to do", regarding the protection, fulfillment, and promotion of human rights, of which non-discrimination is an integral part. In that regard, a state must take necessary measures horizontally, concerning relationships between persons. In other words, States are obliged to take and implement positive administrative and legal measures regarding the protection of human rights. In brief, positive obligations require positive intervention by states, while negative obligations require refraining from intervention (Council of Europe 2008: 11).

The fulfillment of the positive and negative obligations of the state in the face of discriminatory practices caused by AI systems has become a more important and urgent issue in light of current technological developments. Because while digitalization has transformed the production channels of the private sector, it has developed and differentiated the products and services provided to individuals and has also become an indispensable element of the public sector. Considering the prejudice and discriminatory results arising from the development and implementation of the above-mentioned AI systems, the need to reconsider both the "vertical effect" (Gül and Karan 2011: 140) and the "horizontal effect" (Gül and Karan 2011: 140) of human rights in the context of non-discrimination arises.

The black box effect mentioned above of AI systems may cause problems for compliance with and effective enforcement of existing legislation to protect fundamental rights. Moreover, both affected individuals and legal entities and enforcement authorities may not be aware of those negative results that occurred due to discrimination. Additionally, individuals and legal entities may not have effective access to justice in situations, because such biased decisions may negatively affect them (European Commission 2020a, b: 12).

As stated above, existing regulations on the protection of personal data contain provisions against the probable negative consequences of automated decision-making processes, thus AI applications. For instance, data protection regulations require transparency on personal data processing. According to Article 5 titled principles relating to processing of personal data: "(1) Personal data shall be: (a) processed lawfully, fairly and in a transparent manner in relation to the data subject ('lawfulness, fairness and transparency')". Correspondingly, organizations have the obligation to provide information via a privacy notice, for all steps of an automated decision-making process that involve personal data.[18] Besides, GDPR requires specific transparency requirements for automated decisions. For instance (Council of Europe 2018: 42), according to the GDPR, the controller shall provide the data subject with the following information: "*(…) the existence of automated decision-making, including profiling (…) and, at least in those cases, meaningful information about the logic involved, as well as the significance and the envisaged consequences of such*

[18] In addition, GDPR has also other provisions about transparency. See, Article 12 titled transparent information, communication, and modalities for the exercise of the rights of the data subject, and paragraph 2 of Article 88 titled processing in the context of employment.

processing for the data subject". Additionally, the GDPR requires Data Protection Impact Assessment ("DPIA") if a practice is "likely to result in a high risk to the rights and freedoms of natural persons", and stipulates that individual has a "right not to be subject to" certain decisions.

In terms of the national legislation, the Turkish Personal Data Protection Law numbered 6698 ("PDPL") provides a certain level of protection for individuals on that matter. The PDPL provides general principles for processing data protection in Article 4 paragraph 2, such as "*fairness and lawfulness, being relevant, limited and proportionate to the purposes for which they are processed*". Moreover, according to Article 11 of the PDPL, each person has the right to request to the data controller about him/her to object to the occurrence of a result against the person himself/herself by analyzing the data processed solely through automated systems. On the other hand, other rights that are regulated in Article 11[19] may be applicable case by case, as well.

However, it should be reminded that AI systems do not always work based on processing personal data. Additionally, AI systems have their own characteristic elements and risks as explained above; thus, designing, deployment, and use of those systems may require a specific regulation in order to provide full protection to human rights. Therefore, it may be considered necessary to establish a new legislative framework for new obligations and mechanisms regarding the human rights compliance.

3.5 Recent Regulatory Developments on AI in Europe

This chapter focuses briefly on legislative frameworks and approaches on AI of the European Union ("EU") and Council of Europe ("COE") that have shared a series of documentations on AI and its potential legislation in recent years.

---

[19] *See* Article 11 of the PDPL: Each person has the right to request to the data controller about him/her;

a) to learn whether his/her personal data are processed or not,

b) to demand for information as to if his/her personal data have been processed,

c) to learn the purpose of the processing of his/her personal data and whether these personal data are used in compliance with the purpose,

ç) to know the third parties to whom his personal data are transferred in country or abroad,

d) to request the rectification of the incomplete or inaccurate data, if any,

e) to request the erasure or destruction of his/her personal data under the conditions referred to in Article 7,

f) to request reporting of the operations carried out pursuant to sub-paragraphs (d) and (e) to third parties to whom his/her personal data have been transferred,

g) to object to the occurrence of a result against the person himself/herself by analyzing the data processed solely through automated systems, and.

ğ) to claim compensation for the damage arising from the unlawful processing of his/her personal data.

First, the COE has executed various works on the development, design, and deployment of AI, based on COE's standards on human rights, democracy, and the rule of law (Council of Europe 2022). The COE has mostly focused its efforts on AI via the Ad Hoc Committee on Artificial Intelligence[20] ("CAHAI") in recent years. Under the authority of the Committee of Ministers, the CAHAI is instructed to examine the feasibility and potential elements of a legal framework for AI on the basis of COE's standards on human rights, democracy, and the rule of law, and broad multi-stakeholder consultations.

COE's works could be listed on a long line, however, in terms of subject matter, this article will mention only a few of its important works. One of them is Feasibility Report that has been published by CAHAI in 2020. In the Feasibility Report, it is emphasized that a legal response should be developed for the purpose of filling legal gaps in existing legislation and tailored to the specific challenges raised by AI systems (CAHAI 2020a, b: 2). It presents assessments on opportunities and risks arising from the design, development, and application of AI on human rights, rule of law and democracy, and maps instruments that can be applicable for AI (CAHAI 2020a, b: 5–13). In addition, it also examines the main elements of a legal framework for the design, development, and application of AI by focusing on key substantive rights and key obligations (CAHAI 2020a, b: 27–45). Subsequently, it indicates possible options such as modernizing existing binding legal instruments, adoption of a new binding legal instrument and non-binding legal instruments (CAHAI 2020a, b: 45–50), and possible practical and follow-up mechanisms to ensure compliance and effectiveness of the legal framework, for instance, human rights due diligence, certification and quality labeling, audits, regulatory sandboxes, and continuous, automated monitoring (CAHAI 2020a, b: 51–56). On the other hand, the document titled "Towards the Regulation of AI Systems", prepared by the CAHAI Secretariat in 2020 via compilation of contributions, has evaluated the global developments in the regulation of AI within the framework of the Council's standards on human rights, democracy, and rule of law (CAHAI 2020a, b; Çetin and Kumkumoğlu 2021: 34).

In parallel with the COE legislative framework on AI, the EU has also played a pioneer role in this matter as a global standard setter on tech regulation (Council of Europe, Presidency of the Committee of Ministers 2021). Since the foundation of the AI expert group and the European AI alliance in 2018, the EU has published various documentation and constituted organizations in recent years (European Commission). In 2020, in order to set out policy options for supporting a regulatory and investment-oriented approach on AI, a White Paper on Artificial Intelligence a European Approach to Excellence and Trust ("White Paper") has been published by the European Commission. White Paper has mainly dual approaches that promote the uptake of AI and address the risks associated with certain uses of this new technology. It is stated that there is a need to step up action at multiple levels in order to build an ecosystem of excellence and focus on these key issues as follows:

[20] See details on CAHAI: Council of Europe, Ad Hoc Committee on Artificial Intelligence, https://www.coe.int/en/web/artificial-intelligence/cahai, Accessed Feb 1, 2022.

skills, the efforts of the research and innovation community, SMEs, partnership with private sector, adoption of AI by public sector, securing access to data and computing infrastructures, international aspects and working with Member States (European Commission 2020a, b: 5–9).

As a critical piece, it evaluates the regulatory framework by considering human rights from the perspective of AI risks[21], and it makes a list of key features of mandatory legal requirements for the relevant actors, regarding the future regulatory framework for AI. It is indicated that the applicable legal requirements must be complied with in practice and be effectively enforced both by competent national and European authorities and by relevant affected parties (European Commission 2020a, b: 23).

After the White Paper, the EU has published the draft AIA in April 2021 (European Commission 2021; Council of European Union 2021, 2022) that establishes a comprehensive structure for all AI actors via the risk-based approach (European Commission 2021). It is underlined that the risk-based approach includes four risk degrees: unacceptable risk, high risk, limited risk, and minimal risk (European Commission 2021). Whilst the prohibited AI practices are indicated in Article 5, with a few exemptions, high-risk AI practices are classified in Article 6, and these AI systems are referred to in Annex III that shall also be considered high risk.

In this context, the AIA sets out the risk-based requirements for high-risk AI systems and minimum information requirements for certain other AI systems (European Parliament 2021: 4). These requirements shall be mostly applied to providers[22] as they must undergo conformity assessment, via standardization organizations and notified bodies (Veale and Borgesius 2021: 104).

The AIA has been criticized by the law community, civil society, and other actors from various perspectives (Access Now et al. 2021; Smuha et al. 2021; Veale and Borgesius 2021). For instance, in the views of EDRI (EDRI 2021), it is indicated that the AIA may fail to protect people from harmful biometric methods that have the significant threats posed to people's dignity and right to non-discrimination. According to these critiques, such practices can remain harmful results by putting people into boxes of their gender or making guesses about people's future behavior based on predictions about their race or ethnicity.

---

[21] The requirements for high-risk AI applications could consist of the following key features: training data; data and record-keeping; information to be provided; robustness and accuracy; human oversight; specific requirements for certain AI applications, such as those used for purposes of remote biometric identification. See European Commission (2020b).

[22] Provider is defined in Article 3 of AIA, as "*means a natural or legal person, public authority, agency or other body that develops an AI system or that has an AI system developed with a view to placing it on the market or putting it into service under its own name or trademark, whether for payment or free of charge*".

3.6 AI Compliance Mechanisms and AI Governance

In light of the above, the auditability of the fulfillment of the requirements and the related obligations stipulated in the potential regulations regarding AI systems is beneficial and important in terms of reducing and preventing human rights violations, especially non-discrimination. Accordingly, the need for an end-to-end compliance mechanism arises for the creation of transparent and explainable AI systems and for ensuring to deploy and use legal, ethical, and technically robust AI systems. In this context, this chapter will evaluate various compliance mechanisms such as AI impact assessments and audit institutions.

First, the Feasibility Report published by the COE, has set out several compliance mechanisms and it is outlined that each member state should ensure a national regulatory compliance mechanism with any future legal framework (CAHAI 2020a, b: 51). In this context, the report has mentioned different actors as assurers, developers, or operators and users that they have contributions to create a new culture of AI applications, and it has examined which mechanisms can be applied for these actors (CAHAI 2020a, b: 51–53). Afterward, examples of types of compliance mechanisms and the framework's core values[23] should be considered for clarification of these mechanisms.

The mentioned compliance mechanisms have involved human rights due diligence, including human rights impact assessments, certification and quality labeling, audits, regulatory sandboxes, and continuous, automated monitoring. To briefly mention these mechanisms, human rights due diligence is considered a requirement for companies in order to fulfilling their responsibility to respect human rights by highlighting the United Nations Guiding Principles on Business and Human Rights ("UNGPs") (United Nations 2021). It is considered as a mechanism which effectively identifies, mitigates, and addresses risks and impacts. Additionally, it is stated that it should be part of an ongoing assessment process rather than being a static exercise, via a holistic approach which covered all relevant civil, political, social, cultural, and economic rights (CAHAI 2020a, b: 54). In terms of certification and quality labeling (CAHAI 2020a, b: 54), the mechanism can apply to AI embedded products systems or organizations developing or using AI systems and reviewed regularly. This method could be applied voluntarily for systems that pose a low risk or mandatorily for systems that pose higher risks, by depending on the maturity of the ecosystem, via recognized bodies. Thirdly, audit is set forth as a mechanism to verify integrity, impact, robustness, and absence of bias of AI-enabled systems, by exercising throughout the lifecycle of every of them that can negatively affect human rights, democracy, and the rule of law, via by experts or accredited groups (CAHAI 2020a, b: 54). Forth, regulatory sandboxes (CAHAI 2020a, b: 54–55) could ensure a strengthen innovative capacity in the field of AI, in consideration of its agile and safe approach to testing new technologies. Lastly, when AI systems have significant risk, a continuous automated monitoring mechanism is dealt with as a requirement

[23] These values are listed as follow: Dynamic (not static), technology adaptive, differentially accessible, independent, evidence-based. See CAHAI (2020a), p. 53.

to be continually monitored and evaluated AI systems' operation for the purpose of ensuring compliance with established norms (CAHAI 2020a, b: 55).

In parallel to COE's works, the EU has also drawn attention to issues related to compliance mechanisms. As mentioned in the White Paper, the Commission has accentuated a prior conformity assessment for high-risk applications to verify and ensure the mandatory requirements, in general (European Commission 2020a, b: 21). In this context, it is stated that a prior conformity assessment as ex ante[24] could involve certain procedures for testing, inspection or certification, by covering checks of the algorithms and of the datasets used in the development phase (European Commission 2020a, b: 24; Mökander et al. 2021: 248). In parallel to this, the Commission has underlined that *ex-post* controls as a monitoring of compliance should involve AI systems being tested by third parties such as competent authorities (European Commission 2020a, b: 24). However, even if an AI system does not pose high risk, a voluntary labeling option is to be addressed to amplify trust for AI systems (European Commission 2020a, b: 24).

Similar to the approach in the White Paper, the AIA stated that high-risk AI systems[25] could place in the European market so long as subject to compliance with certain mandatory requirements and an ex ante conformity assessment (European Commission 2021). For instance, conformity assessment procedures to be followed for each type of high-risk AI system have been explained extensively in the AIA[26].

On the other hand, as mentioned in the Explanatory Memorandum, transparency and traceability of the AI systems coupled with strong *ex post* controls could amplify effective redress for persons who affected infringements of fundamental rights (European Commission 2021: 11). For instance, as the *ex post* control, the monitoring and reporting obligations for providers have been set, for the purposes of post-market monitoring and investigating on AI-related incidents and malfunctioning. In this context, it is envisaged that market surveillance authorities would be established to control the market and investigate compliance with the obligations and requirements for all high-risk AI systems already placed on the market (European Commission 2021: 15).

To sum up, a need for good governance arises to lead the way in the proper execution of compliance mechanisms and can undertake the task of consulting (depending

---

[24] *Ex ante conformity assessments take place before a system is placed on the market. And post-market monitoring is a type of ex post compliance check.* See footnote 21 at Mökander et al. (2021).

[25] The classification rules and identifies categories of high-risk AI systems are regulated in Chap. 1 of Title III; certain legal requirements for high-risk AI systems in relation to data and data governance, documentation and recording keeping, transparency and provision of information to users, human oversight, robustness, accuracy and security are regulated in Chap. 2; a horizontal obligation for high-risk AI systems is regulated in this chapter; and Chap. 4 sets the framework for notified bodies to be involved as independent third parties in conformity assessment procedures. See details: European Commission (2021), Regulation laying down harmonized rules on artificial intelligence (Artificial Intelligence Act).

[26] See Chap. 5 of AIA involves the conformity assessment procedures to be followed for each type of high-risk AI system.

on the situation), to avoid violation of human rights and failure to fulfill responsibilities, and to ensure sustainability of both. In this context, the AIA sets up the governance systems at the Union and national levels[27]. At the Union level, the proposal establishes a European Artificial Intelligence Board, composed of representatives from the Member States and the Commission. At the national level, Member States will have to designate one or more national competent authorities and, among them, the national supervisory authority, for the purpose of supervising the application and implementation of the regulation (European Commission, Explanatory Memorandum 2021: 15).

In other respects, in the Joint Opinion on AIA of the European Data Protection Board (EDPB) and European Data Protection Supervisory (EDPS), it has been criticized that a predominant role has been assigned to the Commission in the Board; therefore, it has been stated that such role conflicts with the need for an AI European body to be independent of any political influence (European Data Protection Board and European Data Protection Supervisory 2021: 3). Plus, the involvement of independent experts in the development of the EU policy on AI has been recommended (European Data Protection Board and European Data Protection Supervisory 2021: 15).

On the other hand, it could also be taken into consideration to make existing institutions on human rights effective in terms of design, development, deployment, and use of AI. Consequently, the Council of Europe has emphasized the issues related to the prior consultation with Equality Bodies concerning public sector bodies using AI systems. According to this approach (Council of Europe, 2018: 57),

> an Equality Body could help to assess whether training data are biased when public sector bodies plan projects that involve AI decision-making about individuals or groups. Public sector bodies could be required to regularly assess whether their AI systems have discriminatory effects. In a similar vein, Equality Bodies could also require each public sector body using AI decision-making about people to ensure that it has sufficient legal and technical expertise to assess and monitor risks. Besides, Equality Bodies could help to develop a specific method for a human rights and AI impact assessment.

In summary, regarding the compliance mechanisms, it seems that ex ante control is important in terms of determining whether AI systems carry a risk of bias or not, in terms of making these systems a transparent and explainable structure before they are implemented; on the other hand, *ex post* compliance mechanisms also constitute a vital phase, in order to determine AI systems that causes discriminatory results, especially for the learning AI systems that their possible violations cannot be detected with an ex ante control.

[27] See Title VI of the AIA.

3.7 New Procedural Rights to Discuss

New compliance and audit mechanisms become prominent in discussions of combatting discrimination in AI systems, substantial criticisms are also expressed regarding their shortcomings. Smuha et al. (2021) noted on AIA that the enforcement architecture relies heavily on self-conformity assessments[28]. As a consequence of this, it leaves too much discretion to AI providers in assessing risks to fundamental rights without any ex ante review or systematic ex post review, no procedural rights are granted to individuals affected by AI systems, and no complaints mechanism is foreseen.

On the other hand, it is outlined above that, while there are measures aiming to ensure explainability in the data protection framework such as the requirement of privacy notice or the right not to be subject to a decision based solely on automated processing, it is also examined that these measures seem to be insufficient in the context of AI that processes non-personal data (Wachter and Mittelstadt 2019: 55) and has other specific aspects exemplified above. Under these circumstances, along with above discussed "Right to explanation" for AI systems; "The Right to Contest AI" and "Right to be Notified about Discriminative AI" may also be defined as a complimentary procedural right to these compliance mechanisms. However, these possible procedural rights also have their deficiencies.

Firstly, "the right to explanation" has been previously discussed extensively in literature from the perspective of data protection and AI, and it is justifiably described as a hard task in the context of AI and discrimination. One reason is that it is not technically feasible at a great rate (Mittelstadt et al. 2019: 3; Sovrano et al. 2021: 10), and the other reason is legal concepts of non-discrimination cannot be adopted to automated systems in an easy fashion (Wachter et al. 2021). Secondly, creating an effective "the right to contest AI" is a complex problem and there is not a one-size-fits-all solution (Kaminski and Urban 2021). According to Kaminski and Urban (Kaminski and Urban 2021: 2031),

> Designing a successful contestation mechanism requires attention not just to contestation itself, and not just to the algorithm, but to the entire decision-making system—human, machine, and organizational—together with the underlying legal framework.

Thirdly, the "Right to be Notified about Discriminative AI" can be basically defined as a right to know that there is a discrimination. In order to make this right effective, it may be beneficial to propose putting a new obligation on the shoulder of AI provider in case of a detection of a discrimination incident during any phase of AI systems. In this case, the provider should notify the individuals that has been directly or indirectly discriminated. This "Right to be Notified about Discriminative AI" must be backed up with a clear notification criterion, thresholds, and an effective redress mechanism.

---

[28] Moreover, they recommend that the conformity assessment regime should be strengthened with more ex-ante independent control, to ensure their 'trustworthy' status. See Smuha et al., p.37.

3.8 Regulatory Scenery in Turkey

Overall, it may be concluded that a holistic regulatory approach is needed to deal with the non-discrimination and AI problem. In Turkey, there were some early promises on paper about that matter. First, the Ministry of Justice declared "Action Plan on Human Rights" in March 2021 (Turkey Ministry of Justice 2021: 108), One of the goals of the Action Plan was titled "Protection of Human Rights in Digital Environment and Against Artificial Intelligence Applications" and it stipulates these two sub-actions:

> The legislative framework and ethical principles concerning the field of artificial intelligence will be established in consideration of international principles, and measures will be taken regarding the protection of human rights from this aspect.", and "Artificial intelligence applications will be used in the judiciary in conformity with the principles and recommendations of the Council of Europe and without prejudice to the principle of protection of legal guarantees.

Similarly, the National Artificial Intelligence Strategy has been published by the Presidency of the Republic of Turkey—Digital Transformation Office (DDO) in August 2021. In this document, DDO emphasized the importance of the legal and ethical aspects regarding the discrimination and envisioned a holistic regulatory approach and an effective international cooperation and know-how sharing (Turkey Digital Transformation Office 2021: 9). Plus, Turkey was represented and actively participated in CAHAI's agenda and meetings with different shareholders.

3.9 Conclusion

There is a need to adopt a different approach for non-discrimination in this century because anti-discrimination rules and mechanisms cannot completely find a solution to new problems caused by AI. Therefore, it may be necessary to reevaluate existing rules and mechanisms or to establish new rules and mechanisms. We are of the opinion that it would be useful to consider various effects of AI systems on discrimination. Transparency and explainability should be primary purposes for development phases of AI systems, and these principles should also cover deployment phases. To mitigate violation of human rights—especially non-discrimination—compliant mechanisms should be implicated by comprising ex ante and ex post periods. "Right to be Notified about Discriminative AI" that can be basically defined as a right to know that there is a discrimination, should be considered to promote and to protect the respect of human rights. Turkey, which strives to take important steps toward digitalization, should take into account developments in regulation, especially in Europe, about AI, human rights, rule of law and democracy.

References

Access Now et al (2021) An EU artificial intelligence act for fundamental rights a civil society statement. https://www.accessnow.org/cms/assets/uploads/2021/11/joint-statement-EU-AIA.pdf

Australian Government. Australia's AI Ethics Principles. https://www.industry.gov.au/data-and-publications/australias-artificial-intelligence-ethics-framework/australias-ai-ethics-principles

Barocas S, Selbst AD (2016) Big data's disparate impact. Calif Law Rev 671. https://www.californialawreview.org/wp-content/uploads/2016/06/2Barocas-Selbst.pdf

Bathaee Y (2018) The artificial intelligence black box and the failure of intent and causation. Harvard J Law Technol 31(2). https://jolt.law.harvard.edu/assets/articlePDFs/v31/The-Artificial-Intelligence-Black-Box-and-the-Failure-of-Intent-and-Causation-Yavar-Bathaee.pdf

CAHAI (2020a) Feasibility study. https://rm.coe.int/cahai-2020-23-final-eng-feasibility-study-/1680a0c6da

CAHAI (2020b) Towards regulation of AI systems. https://www.rm.coe.int/prems-107320-gbr-2018-compli-cahai-couv-texte-a4-bat-web/1680a0c17a

CAHAI. Members of CAHAI. https://rm.coe.int/list-of-cahai-members-web/16809e7f8d

CAHAI: Council of Europe, Ad Hoc Committee on Artificial Intelligence. https://www.coe.int/en/web/artificial-intelligence/cahai

Casey B, Farhangi A, Vogl R (2019) Rethinking explainable machines: the GDPR's "Right to Explanation" and the rise of algorithmic audits in enterprise. Berkeley Technol Law J. https://ddl.stanford.edu/sites/g/files/sbiybj9456/f/Rethinking%20Explainable%20Machines.pdf

Çetin S, Kumkumoğlu AK (2021) Yapay Zekâ Stratejileri ve Hukuk. On İki Levha Yayıncılık, İstanbul

Chazette L, Brunotte W, Speith T (2021) Exploring explainability: a definition, a model, and a knowledge catalogue. https://arxiv.org/pdf/2108.03012.pdf

European Commission, High Level Expert Group on Artificial Intelligence (2019) Ethics guidelines for artificial intelligence. https://www.aepd.es/sites/default/files/2019-12/ai-ethics-guidelines.pdf

Confalonieri R, Coba L, Wagner B, Besold TR (2020) A historical perspective of explainable artificial intelligence. Wiley Interdiscip Rev. https://doi.org/10.1002/widm.1391#widm1391-bib-0063

Convention against Torture and Other Cruel, Inhuman or Degrading Treatment or Punishment. https://www.ohchr.org/en/professionalinterest/pages/cat.aspx

Convention on the Rights of Persons with Disabilities. https://www.un.org/development/desa/disabilities/convention-on-the-rights-of-persons-with-disabilities/convention-on-the-rights-of-persons-with-disabilities-2.html

Convention on the Rights of the Child. https://www.ohchr.org/en/professionalinterest/pages/crc.aspx

Council of Europe (2008) By Jean-François Akandji-Kombe. Avrupa İnsan Hakları Sözleşmesi Kapsamında Pozitif Yükümlülükler. https://inhak.adalet.gov.tr/Resimler/Dokuman/10122019112811poizitif_yukumluluk.pdf

Council of Europe (2018) Study by Prof. Frederik Zuiderveen Borgesius. Discrimination, artificial intelligence, and algorithmic decision making. https://rm.coe.int/discrimination-artificial-intelligence-and-algorithmic-decision-making/1680925d73

Council of Europe. Presidency of the Committee of Ministers (2021) Human rights in the era of AI: Europe as international standard setter for artificial intelligence. https://www.coe.int/en/web/portal/-/human-rights-in-the-era-of-ai-europe-as-international-standard-setter-for-artificial-intelligence

Council of Europe (2022) Council of Europe's work in progress. https://www.coe.int/en/web/artificial-intelligence/work-in-progress

Council of European Union (2021) Presidency compromise. Brussel. https://data.consilium.europa.eu/doc/document/ST-14278-2021-INIT/en/pdf

Council of European Union (2022) Presidency compromise text. Brussel. https://www.caidp.org/app/download/8369094863/EU-AIA-2021-0106(COD)-12012022.pdf?t=1643032400
Doran D, Schulz S, Besold TR (2017) What does explainable AI really mean? A new conceptualization of perspectives. https://arxiv.org/pdf/1710.00794.pdf
ECHR. European convention on human rights. https://www.echr.coe.int/documents/convention_eng.pdf
EDRI (2021) EU's AI law needs major changes to prevent discrimination and mass surveillance. https://edri.org/our-work/eus-ai-law-needs-major-changes-to-prevent-discrimination-and-mass-surveillance/
Edwards L, Veale M (2017) Slave to the algorithm? Why a 'Right to an Explanation' is probably not the remedy you are looking for. Duke Law Technol Rev 18–84. https://scholarship.law.duke.edu/cgi/viewcontent.cgi?article=1315&context=dltr
European Commission, Eurostat (2007) Handbook on data quality assessment methods and tools. https://unstats.un.org/unsd/dnss/docs-nqaf/Eurostat-HANDBOOK%20ON%20DATA%20QUALITY%20ASSESSMENT%20METHODS%20AND%20TOOLS%20%20I.pdf
European Commission. A European approach to artificial intelligence. https://digital-strategy.ec.europa.eu/en/policies/european-approach-artificial-intelligence
European Commission, High Level Expert Group on Artificial Intelligence (2020a) The assessment list for trustworthy artificial intelligence (ALTAI) for self-assessment. https://futurium.ec.europa.eu/en/european-ai-alliance/pages/altai-assessment-list-trustworthy-artificial-intelligence
European Commission (2020b) White paper on artificial intelligence—a European approach to excellence and trust. https://ec.europa.eu/info/sites/default/files/commission-white-paper-artificial-intelligence-feb2020b_en.pdf
European Parliament (2021) Briefing on artificial intelligence act. https://www.europarl.europa.eu/RegData/etudes/BRIE/2021/694212/EPRS_BRI(2021)694212_EN.pdf
European Union Agency for Fundamental Rights (2019) Data quality and artificial intelligence—mitigating bias and error to protect fundamental rights. https://fra.europa.eu/sites/default/files/fra_uploads/fra-2019-data-quality-and-ai_en.pdf
Federal Trade Commission (2021) Aiming for truth, fairness, and equity in your company's use of AI. https://www.ftc.gov/news-events/blogs/business-blog/2021/04/aiming-truth-fairness-equity-your-companys-use-ai
Felzmann H, Villaronga EF, Lutz C, Larrieux AT (2020) Towards transparency by design for artificial intelligence. Springer. https://www.researchgate.net/publication/345977052_Towards_Transparency_by_Design_for_Artificial_Intelligence
Gohel P, Singh P, Mohanty M (2021). Explainable AI: current status and future directions. IEEE Access. https://arxiv.org/pdf/2107.07045v1.pdf
Government of Canada (2021) Responsible use of artificial intelligence (AI). https://www.canada.ca/en/government/system/digital-government/digital-government-innovations/responsible-use-ai.html#toc1
Gül İI, Karan U (2011) Ayrımcılık Yasağı: Kavram, Hukuk, İzleme ve Belgeleme. İstanbul Bilgi Üniversitesi Yayınları. https://insanhaklarimerkezi.bilgi.edu.tr/media/uploads/2015/02/24/Ayrimcilik_Yasagi_Kavram_Hukuk_Izleme_ve_Belgeleme.pdf
Integrated Innovation Strategy Promotion Council of Japan (2019) Social principles of human-centric AI. https://www.cas.go.jp/jp/seisaku/jinkouchinou/pdf/humancentricai.pdf
International Convention on the Elimination of All Forms of Racial Discrimination https://www.ohchr.org/en/professionalinterest/pages/cerd.aspx
International Covenant on Civil and Political Rights. https://www.ohchr.org/en/professionalinterest/pages/ccpr.aspx
International Covenant on Economic, Social and Cultural Rights. https://www.ohchr.org/en/professionalinterest/pages/cescr.aspx
Istanbul Convention. https://rm.coe.int/168008482e

Kaminski ME, Urban JM (2021) The right to contest AI Columbia Law Review vol 121, no. 7 U of Colorado Law Legal Studies Research Paper No 21–30. SSRN: https://papers.ssrn.com/sol3/papers.cfm?abstract_id=3965041

Karan U (2015) Bireysel Başvuru Kararlarında Ayrımcılık Yasağı ve Eşitlik İlkesi. Anayasa Yargısı 32. https://www.anayasa.gov.tr/media/4440/8.pdf

Kartal E (2021) Makine Öğrenmesi ve Yapay Zekâ. Tıp Bilişimi. pp 360–361. https://cdn.istanbul.edu.tr/file/JTA6CLJ8T5/4270B8B9B702415FA27CAD516B4FF683

Mecham S, Georgia I, Nauck D, Viginas B (2019) Towards explainable AI: design and development for explanation of machine learning predicstions for a patient readmittance medical application. Springer. https://core.ac.uk/download/pdf/222828989.pdf

Mittelstadt B, Russell C, Wachter S (2019) Explaining explanations in AI. https://arxiv.org/pdf/1811.01439.pdf?ref=https://githubhelp.com

Mökander J, Axente M, Casolari F, Floridi L (2021) Conformity assessments and post-market monitoring: a guide to the role of auditing in the proposed European AI regulation. Springer. https://doi.org/10.1007/s11023-021-09577-4

National Institute of Standards and Technology (2021) Four principles for explainable artificial intelligence. https://nvlpubs.nist.gov/nistpubs/ir/2021/NIST.IR.8312.pdf

Nidirect. As and A levels. https://www.nidirect.gov.uk/articles/and-levels#toc-0

OECD (2019) AI principles. https://oecd.ai/en/ai-principles

Porter J (2020) UK ditches exam results generated by biased algorithm after student protests. The Verge. https://www.theverge.com/2020/8/17/21372045/uk-a-level-results-algorithm-biased-coronavirus-covid-19-pandemic-university-applications#:~:text=The%20UK%20has%20said%20that,Reuters%20and%20BBC%20News%20report.&text=In%20the%20UK%2C%20A%2Dlevels,around%20the%20age%20of%2018

Protocol 12 to the ECHR. https://www.echr.coe.int/Documents/Library_Collection_P12_ETS177E_ENG.pdf

Ramcharan BG (1981) Equality and nondiscrimination, the international bill of rights: the covenant on civil and political rights. In: Henkin L (ed) Convention on the elimination of all forms of discrimination against women. Columbia University Press, New York. https://www.ohchr.org/en/professionalinterest/pages/cedaw.aspx

Saeed W, Omlin C (2021) Explainable AI (XAI): a systematic meta-survey of current challenges and future opportunities. https://arxiv.org/pdf/2111.06420.pdf

Smuha N, Ahmed-Rengers E, Harkens A, Li W, MacLaren J, Pisellif R, Yeung K (2021) How the EU can achieve legally trustworthy AI: response to the European commission's proposal for an artificial intelligence act. LEADS Lab University of Birminghom. https://papers.ssrn.com/sol3/papers.cfm?abstract_id=3899991

Sovrano F, Vitali F, Palmirani M (2021) Making things explainable vs explaining: requirements and challenges under the GDPR. https://arxiv.org/pdf/2110.00758.pdf

European Data Protection Board & European Data Protection Supervisory (2021) Joint opinion 5/2021 on the proposal for a regulation of the European parliament and of the council laying down harmonised rules on artificial intelligence (Artificial Intelligence Act). https://edpb.europa.eu/system/files/2021-06/edpb-edps_joint_opinion_ai_regulation_en.pdf

Turkey Digital Transformation Office (2021) National artificial intelligence strategy. https://cbddo.gov.tr/SharedFolderServer/Genel/File/TRNationalAIStrategy2021-2025.pdf

Turkey Ministry of Justice (2021) Action plan on human rights. https://inhak.adalet.gov.tr/Resimler/SayfaDokuman/1262021081047Action_Plan_On_Human_Rights.pdf

Turri V (2022) What is explainable AI? Carnegie Mellon University. SEI Blog. https://insights.sei.cmu.edu/blog/what-is-explainable-ai/

UK Government (2019) Guidance: assessing if artificial intelligence is the right solution. https://www.gov.uk/guidance/assessing-if-artificial-intelligence-is-the-right-solution

Veale M, Borgesius FZ (2021) Demystifying the draft EU artificial intelligence act. Computer Law Review International. https://arxiv.org/ftp/arxiv/papers/2107/2107.03721.pdf

Wachter S, Mittelstadt B (2019) A right to reasonable inferences: re-thinking data protection law in the age of big data and AI. Columbia Bus Law Rev. SSRN: https://papers.ssrn.com/sol3/papers.cfm?abstract_id=3248829

Wachter S, Mittelstadt B, Russell C (2021) Why fairness cannot be automated: Bridging the gap between EU non-discrimination law and AI. Comput Law Secur Rev 41:105567. ISSN 0267-3649. https://doi.org/10.1016/j.clsr.2021.105567

Selin Çetin Kumkumoğlu After studying Japanese in 2012, she started law faculty. She has gained experience in various law firms that provide national and international consultancy on information and technology law. She currently works as a lawyer in the field of information and technology law, especially personal data protection, and artificial intelligence. After her law faculty, she got a master's degree in information and technology law. She worked as a trainee in the European Committee on Legal Co-operation-Council of Europe. At the same time, she is secretary-general of the Information and Technology Law Commission of the Istanbul Bar Association. She leads the Artificial Intelligence Working Group established within the Commission. She is also a member of the Information Technology Commission of the Union of Turkish Bar Associations as the representative of the Istanbul Bar Association. She is a board member of the Understanding Security Alliance Türkiye Chapter.

Ahmet Kemal Kumkumoğlu He graduated from Galatasaray University Law Faculty in 2012 and completed his master's degree in "Law and Digital Technologies" at Leiden University in 2018. He is one of the founding partners of Kumkumoğlu Ergün Cin Özdoğan Attorney Partnership (KECO Legal). Along with his lawyering practice in areas of expertise such as criminal law, IT law, and sports law, he also focuses on the intersection of human rights and digital technologies in his regulatory projects and academic works. In parallel, he actively participates in voluntary works related to his expertise in bar associations and NGOs.

Chapter 4
Regulating AI Against Discrimination: From Data Protection Legislation to AI-Specific Measures

Ahmet Esad Berktaş and Saide Begüm Feyzioğlu

Abstract Various legislation regarding data protection acknowledges the right to protection of personal data as fundamental human right and introduces certain legal obligations to people who have access to personal data to prevent this data to be used without data subject's knowledge and even in some cases their consent. Processing personal data by automated decision-making (ADM) systems bears the risk of discrimination. Especially, when these ADM systems use Artificial Intelligence (AI) and machine learning technologies, natural persons' data may be fed into the system to train the model. Hence, natural persons' personal data constitutes a basis for ADM systems' decisions. Data protection legislation includes certain general principles and measures to prevent misjudgments and discrimination. In the scope of these principles and measures, data processing activity shall be adequate, relevant, and limited in relation to the intended purposes, "privacy by design" and "privacy by default" principles and objection mechanisms regarding negative decisions taken exclusively by ADM systems shall be implemented, accountability and risk-based approach shall be considered. On the other hand, data protection legislation may not be sufficient to eliminate all the risks and threats of AI. Hence, specific regulations, guidelines, and recommendations addressing AI are being drafted.

Keywords Protection of personal data · Artificial intelligence · Discrimination

A. E. Berktaş (✉)
Turkish-German University, Istanbul, Turkey
e-mail: esad@berktas.legal

Berktas Legal, Istanbul, Turkey

S. B. Feyzioğlu
The Ministry of Health, Ankara, Turkey

Acettepe University, Ankara, Turkey

M. Kılıç and S. Bozkuş Kahyaoğlu (eds.), *Algorithmic Discrimination and Ethical Perspective of Artificial Intelligence*, Accounting, Finance, Sustainability, Governance & Fraud: Theory and Application, https://doi.org/10.1007/978-981-99-6327-0_4

4.1 Introduction

A human rights perspective on data processing is important because personal data is protected not only by soft law or ethical rules but concrete legal regulations which have serious consequences in case of breach and also effective remedy mechanisms for data subjects (Albrecht 2016). The United Nations (UN) advocates that a human rights-based approach to data collection is necessary to achieve the Sustainable Development Goals (SDG) which were adopted by the Member States of the UN in 2015. The UN agrees that systematic collection and analysis of data have possible benefits to achieve SDG, and acknowledges that systematic data collection poses risks for data subjects' fundamental human rights. The UN states that designing inclusive and participatory data collection processes, ensuring that different groups are represented in the data set, respecting the privacy of data subjects, providing transparency regarding data collection and analysis processes, and holding data collectors accountable are critical for protecting fundamental human rights and freedoms (Human Rights Council 2018). Another Report by the Human Rights Council (2021) focuses on data processing by AI systems. The Report argues that when AI is using personal information and making decisions affecting individuals, right to privacy along with many other fundamental human rights such as rights to health, education, freedom of movement, freedom of peaceful assembly, freedom of association, and freedom of expression may be violated. According to the Report, AI technologies need to call for adequate safeguards to prevent human rights violations.

Various legislation regarding data protection acknowledges the right to protection of personal data as fundamental human right and introduces certain legal obligations to people who have access to personal data to prevent this data to be used without data subject's knowledge and even in some cases their consent. For example, the European Union's (EU) Regulation 2016/679 (General Data Protection Regulation-GDPR) which entered into force by 2018 acknowledges data protection as a fundamental human right. Recital 2 states that no matter which nationality or residence an individual has, processing of personal data should respect fundamental rights and freedoms, especially the right to protection of personal data. Seeing natural persons not only as data sources but also as individuals who should be able to have control over the fate of their own personal data, these regulations introduce certain legal standards and obligations to the parties who have access to personal data to prevent this data to be used without data subject's knowledge and even in some cases their consent. Moreover, there are independent public authorities and deterrent sanctions to contribute to the consistent application of these regulations.

Similarly, data protection regulations of the Republic of Turkey have adopted a human rights-based perspective. Article 20 of the Constitution of the Republic of Turkey regulates "privacy of private life" and states that every individual has a right to demand their personal data to be protected. This right includes "being informed of", "having access to" and "requesting the correction and deletion of personal data". Also, individuals are granted the right to learn if their personal data are used for relevant purposes. The Law on the Protection of Personal Data numbered 6698 (Data

Protection Law) has been prepared based on Article 20 of the Constitution and came into force on 7 April 2016. The Data Protection Law was prepared with an aim to harmonize Turkish legislation with data protection legislation of the EU. Therefore, the EU's Directive 95/46/EC on data protection has been used as a basis for Turkish Data Protection Law. Pursuant to Article 1 of the Law on the Protection of Personal Data, the purpose of protecting personal data is directly associated with fundamental rights and freedoms of individuals and in particular the right to privacy.

As technology advances, new methods and technologies for data processing emerge. Processing and making sense of big data through Artificial Intelligence (AI), Automated Decision-Making (ADM) systems, and machine learning technologies reinvent the concept of data processing and present new legal, technical, and societal challenges. This paper focuses on data processing through AI and ADM systems and discusses how and to what extent GDPR's data protection legislation regulates these systems. The 11th Development Action Plan of the Presidency, the Judicial Reform Strategy Document of the Ministry of Justice, the Human Rights Action Plan, and the Economic Reform Action Plan state that Turkey is to take necessary steps to further comply with GDPR. Therefore, this paper focuses primarily on the GDPR provisions while elaborating on how existing data protection regulations tackle legal issues related to AI and AMD systems. Last but not least, the paper elaborates on newly emerging AI-specific regulations by providing specific examples.

4.2 Data Processing Through AI and ADM Systems

4.2.1 Definitions of AI and ADM

Definition of AI. AI is rather a new yet popular term which lacks a single, universally accepted definition. AI systems have several distinctive features which differentiate them from other algorithms. These distinctive features include ability to learn independently, gain experience, and improve decision-making processes by coming up with different solutions and acting autonomously (Cerka et al. 2017).

For an AI system to learn independently, gain experience, and improve its decision-making processes, extensive amounts of data need to be collected and studied. To ensure this, first of all, all relevant information, either qualitative or quantitative, is gathered from the outside world by any means possible. Then collected information is converted into digital data, which enables computers to read, analyze, identify patterns, classify, and learn from data to come up with relevant solutions. This process of converting all information into digital data is called "datafication" (Mejias and Couldry 2019; Sadowski 2019; Information Resources Management Association (Ed.) 2021; Lycett 2017). Through datafication processes, even the most qualitative information such as human emotions and gestures can be converted into digital data which then may be fed into AI systems (Mejias and Couldry 2019).

Definition of ADM. Similar to AI, ADM systems do not have one single universal definition as they may come with many different features. An ADM system may or may not use AI and machine learning technologies (Araujo et al. 2020). Definition from the Information Commissioner's Office (ICO) (2021), which is the United Kingdom's independent authority to uphold information rights, defines ADM systems as "*automated individual decision-making is a decision made by automated means without any human involvement*". As can be seen from the definition of the ICO, a characteristic feature of ADM systems is that they involve no human intervention.

However, some other definitions do not require "full automation" feature to consider a system ADM. These definitions state that some ADM systems may involve human interaction at certain points. Therefore, whether it is fully or partially automated, if a decision-making process involves systematic machine-based automation to some extent it may be defined as an ADM system (Information Resources Management Association (Ed.) 2021).

4.2.2 The Point Where AI, ADM, and the Right to Data Protection Intersect

Not all data collected and fed into AI or ADM systems fall into the material scope of the GDPR. If data collected and used by these systems do not contain any information relating to an identified or identifiable natural person or make decisions affecting natural persons, they will not be in the scope of GDPR.

On the contrary, if data collected and analyzed by these systems contain personal data or decisions made by these systems have a potential to have an effect on natural persons then how these systems operate will fall under the scope of the GDPR (Sartor and Lagioia 2020).

It is important to note that today personal data is collected and analyzed via AI and ADM systems more than ever. Today's organizations tend to extract and process as much as data possible (Sadowski 2019). This is because "meaningful data" has proven itself to have high value both for commercial entities and public sector.

Recital 6 of the GDPR also points out the recent paradigm change in sharing, collecting, and analyzing personal data. According to the Recital, due to recent technological advancements, the volume of data shared and collected has increased and new threats against protection of people's privacy and personal data have arisen. Both private entities such as commercial companies and governments have access to vast amount of personal data. Also, natural persons willingly make their personal information online at an escalating rate.

Taking the above points expressed in literature and GDPR into consideration, it can be argued that intersection points of these technologies with data protection regulations will keep expanding consistently and exponentially in the near future.

4.2.3 Benefits and Risks Associated with ADM Systems

There are several benefits of using ADM systems. First of all, ADM systems are being associated with speed, efficiency, accuracy, and progress both in private and public sectors (Waldman 2019; Kaun 2021). Especially, in industries such as health care, banking, engineering, nuclear energy plant operations, building self-driving cars, and security administration, an immense volume of data needs to be analyzed by the decision-maker to make correct decisions. ADM systems may help these industries to evaluate these data and come up with correct decisions under complex situations (Parasuraman and Mouloua 2019; Mökander et al. 2021; Waldman 2019). Moreover, mundane repetitive tasks which normally require time and human resources can be delegated to a simple algorithm which will simply apply certain rules at certain situations (Kaun 2021). Therefore, it would not be wrong to assume that ADM systems powered by AI will be more common soon. Nevertheless, with great power comes great responsibility. ADM systems also have some drawbacks which may cause legal issues.

Over-Dependence on Algorithms and Automation Bias. First of all, as digital data volume increases, humans tend to delegate more responsibilities to computers to automate decision-making processes. This may have negative outcomes such as enabling human errors, weakening the control of humans have over decision-making systems, and undermining human skills and experience (Mökander et al. 2021). Algorithms are assisting humans in critical decision-making contexts at an increasing rate because they are believed to reduce human errors. However, some researches indicate that computers are not always more accurate than human beings. According to a research, participants performed better without the assistance of automated decision systems. Humans who used ADM systems made more "errors of omission" and failed to realize important data if it was not pointed out by the ADM. Also, they made more "errors of commission" meaning that they acted in accordance with ADM system even if the ADM system's decision was contradictory to their previous training and there were other valid data showing that ADM system's decision was wrong (Skitka et al. 1999). A very recent report by the United Nations High Commissioner for Human Rights on the right to privacy in digital age also points out that AI-based decisions are not error-proof and they tend to increase the negative effects of seemingly small error rates. The report states that an analysis conducted on hundreds of AI tools for diagnosing COVID-19 and calculating infection risk shows that none of these AI tools are accurate enough for clinical use (European Commission 2021). Therefore, although one of the main reasons of using ADM systems is to avoid errors, it is critical to remember that these systems are still far from being foolproof.

Invasion of Privacy. Second of all, natural persons' personal data may be fed into the system to train ADM systems. As a result, even if decision-making processes are automated, it is still natural for persons' personal data which constitutes as a basis for the decisions which will be made (Araujo et al. 2020). This may result in third parties having an opportunity to analyze, organize, and evaluate personal data more than ever by using algorithms and automated decision-making systems which may help them profile and rank data subjects without the actual owner of the data, i.e. the data subject even realizing it. Hence, no matter how fully automated, ADM systems may pose a threat to individuals' right to privacy (Castets-Renard 2019).

Black Box Algorithms. Another issue about decision-making systems which pose a great thereat to individuals' fundamental rights is that the algorithms which are used by these systems are rarely transparent. Individuals whose personal data are collected and processed by these systems have no knowledge about how these algorithms work. In some cases, not only data subjects but also data controllers which operate these algorithms may not have an in-depth knowledge about how these algorithms work. (Malgieri and Comandé 2017). AI and ADS systems which use non-transparent algorithms are called "black boxes" (Pasquale 2016). When decisions are taken by black box algorithms subjects of the decision are not presented with a plausible and justifiable reason because even the data scientists behind the systems are unable to comprehend how exactly the algorithm came up with that decision. Not being able to provide reasoning for a decision is harmful for many reasons.

First of all, lack of transparency violates individuals' right to object and demand legal remedies. This is especially harmful when the decision has or may have a direct negative effect on an individual's life (Malgieri and Comandé 2017; Pedreschi et al. 2019). Moreover, black box systems may conduct systematic discrimination as they may have been fed with biased real-life data (Pedreschi et al. 2019). When the past's and today's inequalities and discriminatory behaviors are converted into data and fed into an AI or ADM system, discrimination is reinforced in society through algorithms (Kaun 2021). When the algorithms of these systems lack transparency, it is another challenge to object to decisions taken by them and ask for accountability (European Commission 2021).

Specific Cases of Discrimination and Bias. As indicated by Pedreschi et al. (2019), there are already many controversial cases which shows that decisions taken by non-transparent algorithms are controversial. For example, Correctional Offender Management Profiling for Alternative Sanctions (COMPAS) is an assistive software used in the US which scores a criminal from 1 (lowest risk) to 10 (highest risk) in terms of likelihood of committing a crime one more time in the future. The software also assesses the criminals and classifies them into three groups, namely "low", "medium", and "high" risk of recidivism by using more than one hundred factors such as age, gender, and criminal record. Although "race" or "skin color" are not included in the official list of factors, a study conducted found that the software is biased against black people (Rahman 2020; Angwin et al. 2016).

Another example is Amazon.com Inc.'s recruitment algorithm which has been recently accused of being biased against women. It has been argued that the algorithm is inclined to favor males over females for technical positions. According to a news published on Reuters, the algorithm eliminates resumes which include "feminine words" such as "women's chess club captain", "women's college", etc. The reason for this bias is believed to be that the company uses its own previous recruitment data. Since in the past males were more favorable against females, especially for technical vacancies the company's own dataset is also biased and discriminatory against women (Dastin 2018). As can be seen from these real-life examples, discrimination and bias existing in real life easily leak into computers as computers are fed with real-life examples.

4.3 GDPR Against Discriminatory AI and ADM Systems

Today algorithms govern our everyday lives more than ever. Hence, governing the algorithm is as important as designing algorithms governing our lives (Kuziemski and Misuraca 2020). The GDPR is not enacted for the sole purpose of governing AI and ADM systems. That being said, these systems are means of processing data. Pursuant to Article 1 GDPR regulates the legal aspects related to "the protection of natural persons with regard to the processing of personal data". Therefore, GDPR provisions also cover AI and ADM systems in case these systems process data subjects' personal data. In other words, GDPR provisions are still compatible with AI and ADM applications. For example, GDPR contains general principles, technical and administrative measures, mechanisms to seek legal remedies, and sanctions which may help preventing AI and ADM systems from unlawful data processing and taking discriminatory decisions. Proportionality, accountability, transparency, data minimization, the concepts of "privacy by design" and "privacy by default" and risk-based approach are among these principles and measures. Moreover, heavy penalties in case of non-compliance with provisions of the GDPR are also applicable for AI and ADM systems. Hence, GDPR is still a deterrent legal document for processing data through AI and ADM applications. Below are some GDPR provisions which specifically apply to AI and ADM applications.

Transparency, Accessibility, and Right to Object. Pursuant to Article 5 AI and ADM systems are under the obligation to process data in a transparent manner. Pursuant to Article 12 data controller shall take appropriate measures to provide information regarding data processing activities. Pursuant to Article 15 data subject has a right to learn if his or her data is being processed by a data controller and demand information related to this data. Last but not least as per Article 21 data subject has a right to object when a data processing activity influences his or her situation. This right to object includes cases where data controller performs profiling activities which will be further explained below.

In light of these provisions, if AI and ADM systems use black box algorithms, meaning that the reasoning behind these systems is not comprehendible for the data subject or in some cases even unknown to the developer of these systems, Articles 5, 12, 15 and 21 are being violated. Since AI and ADM systems are highly technical, informing the data subject about the details of data processing is not an easy task as the data subject does not always have advanced technical knowledge (Pasquale 2016). In other words, how these principles of transparency, accessibility, and right to object should be realized for AI and ADM systems is not clear in GDPR and needs further elaboration (Wachter et al. 2017).

Profiling and Automated Processing. Pursuant to Article 22 of the GDPR "automated individual decision-making" is "*a decision based solely on automated processing, including profiling, which produces legal effects concerning him or her or similarly significantly affects him or her*". According to Recital 71, a data subject should have a right not to be subject to an evaluation based only on automated processing if that evaluation will have legal or significant legal effects. In other words, if it will include any measure, evaluate personal data aspects, and produce legal or similar significant effects on a data subject, it is the data subject's right to refuse a decision made without human intervention. The Recital includes several examples to further elaborate on what automated individual decision-making is. For example, profiling individuals to make predictions about them with regard to their economic situation, health or reliability is off the limits if it does not involve any human intervention. Similarly, profiling is also defined in Article 4 of the GDPR as "*any form of automated processing of personal data consisting of the use of personal data to evaluate certain personal aspects relating to a natural person, in particular to analyse or predict aspects concerning that natural person's performance at work, economic situation, health, personal preferences, interests, reliability, behaviour, location or movements*".

Both Article 22 and Recital 71 of the GDPR name some exceptional cases where automated decision-making is allowed. These cases are (i) existence of a necessity for a contract between a data subject and a data controller, (ii) Union's or Member State's authorization by law which also includes necessary safeguards to protect data subject and lastly, and (iii) data subject's explicit consent. However, as Article 22 states, even in cases (i) and (iii), data subjects should have a right to demand human intervention and refuse to be evaluated only by ADM systems. Moreover, in these cases, data subjects should also have a right to express their own opinion regarding the decision of the automated system and object to the decision if they prefer to. In other words, if a decision will have legally binding or significant effects on a natural person, who is protected under GDPR, that person has a right to demand not to be subjected to a decision which is taken without any human interference and only by automated processing (European Commission 2021).

European Commission's Article 29 Data Protection Working Party was established under Article 29 of Directive 95/46/EC. It was an advisory body on data protection and privacy until the European Data Protection Board (EDPB) was established after GDPR entered into force. Working Party's "Guidelines on Automated Individual Decision-Making and Profiling for the Purposes of Regulation 2016/679" (2018) elaborates on the scope of automated processing and states that automated individual decision-making, and profiling is a broader concept which means making decisions by using technology and without any human interference. These systems can use any sort of data including; data directly collected from individuals via means such as a questionnaire, data observed about individuals such as GPS data, or data derived from previously created profiling activity such as credit score. However, as explained in the Guideline, GDPR's Article 22 only covers automated decision-making when the decision has legal or similar serious potential effects on individuals. In other words, although automated decision-making may be a broader term, GDPR only regulates when the potential impacts of the decision bear a risk of harming the data subject. This risk of harm may present itself in different ways. For example, a decision which may have a "prolonged or permanent impact on the data subject" or "lead to the exclusion or discrimination" is considered within the scope of Article 22 of the GDPR.

To sum up, according to Article 22 of the GDPR if a data subject is profiled or subjected to automated decision making he or she has the following rights; (i) the right to an explanation, (ii) right to human intervention, and (iii) right to object. All these rights are applicable to data processed via AI and ADM systems. Nevertheless, as GDPR is not specifically designed for AI or ADM systems, the article does not contain specific details about how to realize these rights.

Privacy by Design and Privacy by Default. Article 25 of the GDPR regulates the data controller's obligation to implement the appropriate technical and organizational measures from the beginning, i.e. during the design phase of the data processing in an effective manner. Data controller needs to integrate the necessary safeguards into data processing to protect the rights of data subjects. Among these measures, Article 25 names pseudonymization and data-minimization. Both pseudonymization and data-minimization are measures which can be taken for data which will be fed into AI and ADM systems. Pseudonymization means keeping any information which may lead to the identity of a data subject separately to make sure that data is not easily attributed to that person. Data minimization measure requires data controller to collect and process as little data as possible. Moreover, pursuant to Article 25 (2) of the GDPR data controller is expected to take appropriate technical and organizational measures to ensure that "just enough" data is processed for "just enough" period of time and is accessed by "just enough" number of people.

4.3.1 Specific Regulations on AI and ADM

Although GDPR is applicable to AI and ADM, it needs to be acknowledged that AI and ADM systems are very complex, use a novel technology which may be a "black box" even for its creators, and possibly have a wide-scale effect both on individuals and society in general. Therefore, we shall accept that GDPR cannot answer all legal disputes arising from AI and ADM applications. Unanswered questions may leave technology companies without sufficient guidance. This may cause data protection authorities apply stiff sanctions to these companies and consequently hinder development of AI and ADM technologies. Therefore, enacting specific laws for these novel technologies may be beneficial both for data subjects and data controllers (Sartor and Lagioia 2020).

Indeed, to eliminate the risks and threats of these systems, specific regulations are being drafted. Among the main ethical, regulatory and policy initiatives undertaken at global and European levels in the recent years are the Council of Europe's Recommendation on Human Rights Impact of Algorithmic Systems, UNESCO's Regulation on Ethics of Artificial Intelligence, Council of Europe's Ad hoc committee on AI "CAHAI" which was established in 2019 with an aim to study a prospective legal framework artificial intelligence and last, but not least the Draft EU Act on Artificial Intelligence.

4.3.2 CoE's Recommendation on Human Rights Impacts of Algorithmic Systems

Council of Europe's Recommendation CM/Rec (2020) 1 of the Committee of Ministers to member States on the human rights impacts of algorithmic systems was adopted on 8 April 2020 (Council of Europe Committee of Ministers 2020). The Recommendation does not use the term AI, but prefers using a broader term which is "algorithmic system". The Recommendation defines "algorithmic systems" as "*applications that, often using mathematical optimisation techniques, perform one or more tasks such as gathering, combining, cleaning, sorting, classifying and inferring data, as well as selection, prioritisation, the making of recommendations and decision making*". This definition is a broad one which can easily be adopted to evolving technologies. The recommendation targets not only states but also all public and private institutions. It invites states to enact policies and regulations applicable to design, development, and deployment of algorithmic systems to ensure transparency, accountability, and inclusivity. Moreover, the recommendation underlines the importance of assessing algorithmic systems' impacts on human rights on a regular basis.

4.3.3 UNESCO's Regulation on Ethics of Artificial Intelligence

UNESCO's Regulation on Ethics of Artificial Intelligence was adopted on 24 November 2021 (UNESCO 2021). As expressed under the title "Final Provisions", the regulation is of recommendatory and complementary nature. It does not replace or alter any legal obligation and it is not binding. Nevertheless, the regulation lays out recommendations on how to use AI systems in a responsible manner and encourages all member states to engage all stakeholders including the private sector to take the necessary measures to implement these recommendations. Therefore, these recommendations interest not only governments but also private sector. The regulation clearly states that it is not intended to give a clear definition of AI as it is a technology and concept constantly evolving. As explained under the title "Scope of Application", AI systems process information, integrate models and algorithms, and have a capacity to learn and act autonomously at a varying degree. AI systems may use several methods including, but not limited to machine learning, deep learning, reinforcement learning, etc. Although it is not a binding document and does not create any legal obligations for states or companies, the document is important as it provides a written basis for ethical impact assessment, ethical governance, and risk assessment of AI.

4.3.4 CoE's Ad Hoc Committee on Artificial Intelligence (CAHAI)

CAHAI is an inter-governmental committee consisting of representatives from the Council of Europe's 47 member states and representatives from 6 non-member states as observers. Representatives from other international and regional organizations such as the UNESCO, the European Union, and the OECD also participate in the meetings. Representatives from private sector have concluded exchange of letters and representatives from civil society, research, and academic institutions have also been accepted as observers. Therefore, CAHAI has been established in an inclusive mindset to gather ideas from as many stakeholders as possible. CAHAI is expected to "examine the feasibility and potential elements on the basis of broad multi-stakeholder consultations, of a legal framework for the development, design and application of artificial intelligence". Similar to the Council of Europe's Recommendation on Human Rights Impacts of Algorithmic Systems and UNESCO's Regulation on Ethics of Artificial Intelligence, CAHAI underlines that legal framework for AI should be based on human rights (Council of Europe 2020).

4.3.5 Proposal for Artificial Intelligence Act

In April 2021 the European Commission made a proposal for an EU regulatory framework on AI. The draft is a first attempt to regulate AI systems with a binding legal document. The draft adopts a risk-based approach instead of human rights-based approach and categorizes AI applications into groups as "unacceptable-risk", "high-risk", and "minimal-risk" systems. Target is narrower compared to the above initiatives. AI applications which are not classified as risky are out of scope of this draft. Therefore, it has a narrower scope compared to the advisory documents explained above (European Parliament 2021).

4.4 Conclusion

AI and ADM technologies can be powerful tools to help societies tackle the biggest challenges of our century. On the other hand, potential negative effects of these systems may also result in catastrophic human rights violations today and in the future. Therefore, it is important to set out clear-cut rules and determine appropriate, effective, and concrete safeguards (European Commission 2021). Existing regulations on data protection may also cover AI and ADM applications under their scope because these systems rely heavily on big data which may also contain data related to natural persons. However, as the existing data protection regulations have not been designed by keeping specific problems which may be caused by these systems in mind, in some cases they may not be clear enough or have shortcomings. Therefore, specific legal regulations are needed to regulate these areas (Wachter et al. 2017).

Keeping the rules of the game clear is also an advantage for tech companies which make financial investments in AI and ADM Technologies. This is because data protection authorities have the authority to impose harsh economic sanctions on data controllers. If tech companies have clear-cut rules on how to design and use these systems, they may be more eager to invest and increase their research and development expenditures. As a result, current deficiencies and fallacies in AI and ADM may be overcome eventually. Hence, having more clear-cut rules on AI and ADM systems not only protects human rights but also protects tech companies in the long term by offering more clarity and predictability which may generate positive outcomes for society and individuals by increasing research and development by private companies which may help eliminating harmful effects of AI and ADM systems (Bodea et al. 2018). Indeed, to regulate these systems specific regulations on algorithmic applications are being drafted and special task forces are being formed. These documents and working groups tend to give broad definitions for AI considering that it is a fast-evolving area.

References

Albrecht JP (2016) How the GDPR will change the world. Eur Data Prot L Rev, 2

Angwin J, Larson J, Mattu S, Kirchner L (2016) Machine bias. ProPublica. https://www.propublica.org/article/machine-bias-risk-assessments-in-criminal-sentencing

Araujo T, Helberger N, Kruikemeier S, Vreese CH (2020) In AI we trust? Perceptions about automated decision-making by artificial intelligence. AI Soc, 611–623

Bodea G, Karanikolova K, Mulligan DK, Makagon J (2018) Automated decision-making on the basis of personal data that has been transferred from the EU to companies certified under the EU-U.S. Privacy Shield: fact-finding and assessment of safeguards provided by U.S. law. European Commission

Castets-Renard C (2019) Accountability of algorithms in the GDPR and beyond: a European legal framework on automated decision-making. Fordham Intell Prop Media Ent LJ 30

Cerka P, Grigiene J, Sirbikyte G (2017) Is it possible to grant legal personality to artificial intelligence software systems? Computer Law Sec Rev, 685–699. https://doi.org/10.1016/j.clsr.2017.03.022

Council of Europe (2020) Consultation on the elements of a legal framework on AI. coe.int. https://www.coe.int/en/web/artificial-intelligence/cahai#{%2266693418%22:[1]}
Council of Europe Committee of Ministers (2020) Recommendation CM/Rec (2020) 1 of the Committee of Ministers to member States on the human rights impacts of algorithmic systems. Council of Europe. https://rm.coe.int/09000016809e1154
Dastin J (2018) reuters.com. Amaon scraps secret AI recruiting tool that showed bias against women. https://www.reuters.com/article/us-amazon-com-jobs-automation-insight-idUSKCN1MK08G
European Commission (2021) Are there restrictions on the use of automated decision-making? ec.europa.eu. https://ec.europa.eu/info/law/law-topic/data-protection/reform/rules-business-and-organisations/dealing-citizens/are-there-restrictions-use-automated-decision-making_en#references
European Commission Article 29 Data Protection Working Party (2018) Guidelines on automated individual decision-making and Profiling for the purposes of Regulation 2016/679. European Commission, Belgium
European Parliament (2021) Artificial intelligence act. European parliament. https://www.europarl.europa.eu/RegData/etudes/BRIE/2021/698792/EPRS_BRI(2021)698792_EN.pdf
Human Rights Council (2018) Report of the office of the united nations high commissioner for human rights 'A Human Rights-Based Approach to Data'. United Nations. https://www.ohchr.org/Documents/Issues/HRIndicators/GuidanceNoteonApproachtoData.pdf
Human Rights Council (2021) Report of the office of the united nations high commissioner for human rights 'The right to privacy in the digital age'. United Nations
Information Commissioner's Office (2021) Rights related to automated decision making including profiling. ico.org.uk. https://ico.org.uk/for-organisations/guide-to-data-protection/guide-to-the-general-data-protection-regulation-gdpr/individual-rights/rights-related-to-automated-decision-making-including-profiling/#ib2
Information Resources Management Association (Ed.) (2021) Research anthology on decision support systems and decision management in healthcare, business, and engineering. IGI Global
Kaun A (2021) Suing the algorithm: the mundanization of automated decision-making in public services through litigation. Inf Commun Soc. https://doi.org/10.1080/1369118X.2021.1924827
Kuziemski M, Misuraca G (2020) AI governance in the public sector: three tales from the frontiers of automated decision-making in democratic settings. Telecommun Policy 44(6). https://doi.org/10.1016/j.telpol.2020.101976
Lycett M (2017) 'Datafication': making sense of (big) data in a complex world. Eur J Inf Syst, 381–386.https://doi.org/10.1057/ejis.2013.10
Malgieri G, Comandé G (2017) Why a right to legibility of automated decision-making exists in the general data protection regulation. Int Data Priv Law 7(4):243–265. https://doi.org/10.1093/idpl/ipx019
Mejias UA, Couldry N (2019) Datafication. Internet Policy Rev 8(4)
Mökander J, Morley J, Taddeo M, Floridi L (2021) Ethics-based auditing of automated decision-making systems: nature, scope, and limitations. Sci Eng Ethics 27(44). https://doi.org/10.1007/s11948-021-00319-4
Parasuraman R, Mouloua M (2019) Automation and human performance: theory and applications (Human Factors in Transportation), 1 ed. CRC Press. ISBN 9780367448554
Pasquale F (2016) The black box society: the secret algorithms that control money and information. Harvard University Press. 9780674970847
Pedreschi D, Giannotti F, Guidotti R, Monreale A, Ruggieri S, Turini F (2019) Meaningful explanations of black box AI decision systems. In: Proceedings of the AAAI conference on artificial intelligence, vol 33(1), pp 9780–9784. https://doi.org/10.1609/aaai.v33i01.3301978
Rahman F (2020) COMPAS case study: fairness of a machine learning model. Towards data science. https://towardsdatascience.com/compas-case-study-fairness-of-a-machine-learning-model-f0f804108751
Sadowski J (2019) When data is capital: datafication, accumulation, and extraction. Big Data Soc. https://doi.org/10.1177/2053951718820549

Sartor G, Lagioia F (2020) The impact of the General Data Protection Regulation (GDPR) on artificial intelligence. EPRS | European parliamentary research service. European Union, Brussels. https://doi.org/10.2861/293
Skitka LJ, Mosier KL, Burdick M (1999) Does automation bias decision-making? Int J Hum Comput Stud, 991–1006. https://doi.org/10.1006/ijhc.1999.0252
UNESCO (2021) Recommendation on the ethics of artificial intelligence. UNESCO, Paris. https://unesdoc.unesco.org/ark:/48223/pf0000380455
Wachter S, Mittelstadt B, Floridi L (2017) Why a right to explanation of automated decision-making does not exist in the general data protection regulation. Int Data Priv Law 7(2):76–99. https://doi.org/10.1093/idpl/ipx005
Waldman AE (2019) Power, process and automated decision-making. Fordham L Rev 88(613)

Ahmet Esad Berktaş Works as a lawyer in Istanbul. He completed his LL.B. at Başkent University and his LL.M. in Information Technology and Commerce Law at the University of Southampton. He is a Ph.D. candidate at Turkish-German University and is writing his dissertation on the protection of personal data in the health domain. He worked as a Consultant on Data Protection Law at the Turkish Ministry of Health between 2016 and 2023. He established the Data Protection Unit in the Turkish Ministry of Health just before the Personal Data Protection Law entered into force. He provided legal counseling services to various projects of the World Bank and the World Health Organization and took part as a consultant in several EU projects. His work focuses on data protection law as well as health informatics law and he has many books and papers written on these topics. He teaches data protection law courses at undergraduate and postgraduate degrees at various universities.

Saide Begüm Feyzioğlu She graduated from Koç University Faculty of Law in 2015. After completing her master's degree in the field of "Sustainable Development" at the Blekinge Institute with a scholarship from the Swedish Institute, she worked as a legal consultant and self-employed lawyer in private health institutions between 2018 and 2021. Feyzioğlu has been working as a Health Informatics Law Consultant at the Ministry of Health since 2021.

Chapter 5
Can the Right to Explanation in GDPR Be a Remedy for Algorithmic Discrimination?

Tamer Soysal

Abstract Since the birth of computation with Alan Turing, a kind of "excellence/extraordinary" and "objectivity" has been attributed to algorithmic decision-making processes. However, increasing research in recent years has revealed that algorithms and machine learning systems can contain disturbing levels of bias and discrimination. Today, efforts have accelerated to include "fairness", "transparency", and "accountability" features of algorithms. In this paper, in this new environment created by algorithms, whether the "right to explain" regulation in the GDPR, which entered into force on April 25, 2018, in the EU, can be used as a remedy and its limits will be discussed.

Keywords Algorithms · Discrimination · Opacity · Artificial intelligence · Automated processing · Profiling · GDPR · Right to explanation · GDPR Article 22

5.1 Introduction: Algorithms and Risks in General

An algorithm is "a finite set of precise instructions for performing a computation or for solving a problem (Danesi 2021: 171). An algorithm is a set of instructions for how a computer should accomplish a particular task. The first algorithm in the history of mathematics is considered to be the Euclidean algorithm, which allows us to find the greatest common divisor of two integers. The term algorithm is derived from the word "Alkhorismi", which is the Latin equivalent of the name of the Muslim scholar Al-Harizmi. In his book titled as "Kitâbü'l-Muhtaşar fî hisâbi'l-cebr ve'l-mukâbele", Al-Harizmi created algorithms that develop solutions in first- and second-degree equations in mathematics. The name "algebra" was started to be used based on this book.

T. Soysal (✉)
EU Project Implementation Department, Ministry of Justice, Ankara, Turkey
e-mail: tamer.soysal@adalet.gov.tr

M. Kılıç and S. Bozkuş Kahyaoğlu (eds.), *Algorithmic Discrimination and Ethical Perspective of Artificial Intelligence*, Accounting, Finance, Sustainability, Governance & Fraud: Theory and Application, https://doi.org/10.1007/978-981-99-6327-0_5

In the first step of creating the algorithm, a flowchart is created that shows the steps needed to complete the task. Each step in the flowchart presents a choice, enabling the decision to be reached. This simple flowchart is then converted into computer programs. Each algorithm has certain conditions and presuppositions. Otherwise, machine language translation of tasks will not be possible (Danesi 2021: 173). Algorithms are an abstract, formalized, description of a computational procedure. The output of this procedure is generally referred to as the "decision". Algorithmic decision-making refers to the process by which an algorithm produces the output (Borgesius 2020: 1573).

Since algorithms use a large number of variables and possibilities, machine learning systems containing artificial intelligence have also started to be used frequently in algorithms. Machine learning generally refers to the automated processes of discovering correlations (relationships or patterns) between variables in a dataset to make predictions. Due to the huge amount of data available today, machine learning has been widely used in the last decade (Borgesius 2020: 1574).

Algorithms make big data usable. This is achieved in three stages:

i. Data collection and aggregation of datasets, (ii) Analysis of data, (iii) Actual use of data by applying it to the model (Janssen 2019: 13).

Different techniques are used to analyze the data. Algorithms derive patterns from large datasets by using data mining. Then, four methods are used as classification, clustering, regression, and association techniques. Classification aims to categorize data. Algorithms learn from previously classified examples by systematically comparing different categories. Algorithms can distill rules and apply them to new situations. Clustering techniques aim to group data that is very similar to each other. For example, purchasing behaviors of different customers are collected and a clustering can be made according to them. In classification method, there are predefined classes. On the other hand, in clustering, a new classification is made according to common features, thanks to data analysis.

Regression techniques aim to formulate numerical estimates based on defined correlations derived from datasets. For example, the credit application evaluations of banks include regression techniques. Correlation techniques, on the other hand, try to reveal correlations by establishing connections between data. The movie recommendations of Netflix exemplify attribution techniques (Janssen 2019: 13).

The general view that the equilibrium in the market will be ensured through the price mechanism during the regulation of economic life is expressed with the metaphor of "*invisible hand*" (Minowitz 2004: pp. 381–412), used by the famous economist Adam Smith for once in his books "The Theory of Moral Sentiments" written in 1759 and "The Wealth of Nations" written in 1776 (Smith 2022).[1] Today, the analogy of "*invisible hand*" that regulates the markets is also made for algorithms

[1] "It is not from the benevolence of the butcher, the brewer, or the baker, that we expect our dinner, but from their regard for their own interest…. Each participant in a competitive economy is" led by an invisible hand to promote an end which was no part of his intention". Adam Smith, The Wealth of Nations, Book 1, Chapter 2, https://geolib.com/smith.adam/won1-02.html (last visited, March 1, 2022).

(Engle 2016). Algorithms are useful tools used to perform defined tasks and are often invisible, so they are also referred to as invisible aids (Rainie and Anderson 2017). Wherever there are computer programs, the internet, social media, and smartphone apps; algorithms are also found. All online dating apps, travel websites, GPS mapping systems, voice command systems, face identification, photo classifications work thanks to algorithms. However, this invisible side of algorithms has also raised some concerns in recent years (Rainie and Anderson 2017). Thanks to machine learning systems, Twitter's chatbot named "Tay", which was created for chat in 2016, started to send racist and sexist tweets on Twitter in a short time, as well as insults and political comments.[2] It is frequently stated that speculative movements are created due to automatic purchases, based on algorithms created in the financial system. For example, on 7 October 2016, the British pound depreciated by more than 6% in seconds, partly due to purchases triggered by algorithms.[3] In a study commissioned by the U.S. Senate, it was found that a single visit to a popular news website initiated automatic activity on more than 350 web servers. Information such as identification and tracking of visitors, interests, digital profiling, online behavior patterns, including the delivery of advertisements, is distinguished as processed data. China established a nationwide social credit system and began to base its decisions on public services (Mac Sithigh and Siems 2019). For example, 9 million people with low scores were prevented from purchasing tickets for domestic flights (Janssen 2019: 4).

A new understanding of "algocratic governance", replacing "bureaucratic oligarchies" which was the fashionable concept of a period, is mentioned (Aneesh 2002). It is emphasized that the new algocratic governance creates a third dimension to the existing bureaucratic and panoptic management systems (Aneesh 2002). It is stated that the hierarchy in bureaucratic structures will be replaced by the codes and algorithms; the chain of command will be replaced by the programmability, and the vertical organizations will be replaced by the horizontal organizations.

In his book titled "The Black Box Society: The Secret Algorithms That Control Money and Information" and written in 2015, Lawyer Frank Pasquale, lecturer at the University of Maryland, (Pasquale 2015) defines the society, which is enslaved by invisible algorithms, as the black box society. Black-box algorithms can be used to infer a data subject's location, age, medical condition, political opinions, and other personal information. It is not clear which bits the data algorithms choose and how the algorithms use those bits to provide output. This highlights the "opaque" nature of algorithmic decision-making processes. For example, no user knows how search engine algorithms work. However, search engines access and process a lot of personal data. The data obtained from the algorithmic profiling activities, which is used for employment, credit eligibility assessment, hospital and pharmacy, can be used automatically for many decisions about the person. We have scarcely any

---

[2] Tay, Microsoft's AI chatbot, gets a crash course in racism from Twitter, March 24, 2016, https://www.theguardian.com/technology/2016/mar/24/tay-microsofts-ai-chatbot-gets-a-crash-course-in-racism-from-twitter (last visited March 1, 2022).

[3] The Sterling 'flash event' of 7 October 2016, Markets Committee, https://www.bis.org/publ/mktc09.pdf (last visited, March 1, 2022).

sources of information on the collection and profiling of such data. Such "black box decisions" cannot be questioned in a transparent and reliable manner. Individuals have a right to know how algorithmic decision-making models affect them. However, this opacity in algorithmic decision-making processes, or the "black box" quality as Pasquale describes it, prevents this. Opacity can occur in three different ways (Janssen 2019: 14–15).

i. *Intrinsic opacity*: Intrinsic opacity refers to the opaque nature of algorithms. The faster the machine is capable of learning, the harder it is to understand the reasons behind the decisions.
ii. *Illiterate opacity*: Algorithms are technically complex. Understanding algorithms requires training beyond computer literacy. Most of the people don't know the basic principles that algorithms work and they can't make sense of the codes.
iii. *Intentional opacity*: The most dangerous opacity in terms of discrimination is this kind of opacity. Companies don't want people to know how their algorithms work. This can be done for purposes such as trade secrets, or it can be done so that the situations such as discrimination are not understood.

Frank Pasquale proposed to add four new robotic laws (Pasquale 2020) next to Isaac Asimov's three basic robotic laws (Soysal 2020: 245–325)[4] in order to eliminate this situation. Accordingly:

1. Digital Technologies ought to "complement professionals, not replace them".
2. AI and robotic systems "should not counterfeit humanity".
3. AI should be prevented from intensifying "zero-sum arms races".
4. Robotics and AI systems need to be forced to "indicate the identify of their creators, controllers and owners".

Our bias are always present. It is a fact that in the field of criminal justice, certain bias have been evident in many countries for a long time. For example, when an African–American named Duane Buck, who was sentenced to death for murder in Texas/USA in 1997, was brought before the jury; the jury would decide whether the convict—who killed two people—would be sentenced to life imprisonment with parole or the death penalty. Before the jury trial, a psychologist named Walter Ouijano was called to the hearing and his opinion was asked. When the psychologist, who studied the recidivism rates in Texas prisons, referred to the race of the convict; the Prosecutor on the case asked as follows: "Are you suggesting that, for various complex reasons, the racial factor, namely the fact that the convict is black, increases

[4] In his book titled "I Robot" and written in 1942, Isaac Asimov proposed the following three basic laws:

1. A robot may not injure a human being or, through inaction, allow a human being to come to harm. 2. A robot must obey the orders given it by human beings except where such orders would conflict with the First Law. 3. A robot must protect its own existence as long as such protection does not conflict with the First or Second Law.

the risk of committing crimes in the future? Is this true?" "Yes," the psychologist answered without hesitation. The Jury sentenced Buck to death penalty (O'Neil 2016: 35).

Whether or not the issue of race is brought up openly in the courts in this way, the issue of race has been an important factor in some countries, especially in the USA. In a study carried out in the University of Maryland, it was found that prosecutors were three times more likely to seek the death penalty for African–Americans than for whites convicted of the same crime; and for Hispanics, it was four times more. A study commissioned by the American Civil Liberties Union found that the punishments for blacks in the US federal system were 20% longer than those for whites convicted of similar crimes. Again, the share of African-Americans in the population is 13%, while their rate in American prisons is 40% (O'Neil 2016: 36).

In short, bias has always existed in the field of criminal justice. In this environment, it is possible to think that computer-based risk models, fed and developed with data, will reduce the role of bias in the punishments and a more objective evaluation will be made. However, it should not be forgotten that the effectiveness and usefulness of these models depends on a number of assumptions. Many assumptions play a role in these algorithms, from "how many previous convictions" to the ethnic identity of the person.

As a result of such workshops, I think that we should try to answer the following question:

> Can we eliminate human bias with these new computer-based, artificial intelligence supported models? Or are we camouflaging our current bias with technology?

In many technical fields such as banking and criminal justice, algorithms are extremely complex and mathematical. Many assumptions are embedded in these models, sometimes with biases. Therefore, one of the most important aspects of algorithmic discrimination is this incomprehensible phenomenon called "black box". It is not possible for the average person to interpret and understand these algorithms.

After trying to put forward a general concept in this way, based on one of the many proposals made to solve these dilemmas; I will try to form an idea toward whether the EU's General Data Protection Regulation, which entered into force on 25 May 2018, will be an appropriate instrument, and how much the regulation known as "Right to Explanation" will work.

5.2 EU's General Data Protection Regulation and "Right to Explanation" Concept

The protection of personal data in the EU was regulated by the Directive numbered 95/46 on the Protection of Individuals with regard to the Processing of Personal Data and on the Free Movement of Such Data, which was adopted on 20 February 1995 and entered into force in 1998.[5] And then, as of 25 May 2018, the General Data Protection Regulation (GDPR) was entered into force. With GDPR,[6] the personal data is regulated more broadly than the Directive numbered 95/46. In GDPR, the personal data is defined as any information relating to an identified or identifiable natural person. In the Directive numbered 95/46, on the other hand, the personal data subject was defined as limited to the person and personality characteristics (Directive No. 95/46, Article 2/1-a). Within the scope of GDPR, one or more factors specific to the person's name, identity number, location data, online identifier or the physical, physiological, genetic, spiritual, economic, cultural or social identity of the natural person in question are also specified in this context (GDPR, Article 4/1). Data used in the form of pseudonyms are also included in the scope of protectable personal data.

With Article 5 of the GDPR, the basic principles regarding the processing of personal data were determined. One of the most important of these basic principles is lawfulness, fairness and transparency. Therefore, this principle should always be taken into account in the interpretations of the GDPR.

There is no explicit and individual "right to explanation" regulation in the GDPR text or in the previous regulation in this field, the Directive numbered 95/46. However, when the provisions 13/2-f and 14/2-g of the GDPR regulating the rights of the data subject, the provision 15/1-h on the data subject's right to access personal data and the provisions of Article 22 in general are read and considered with Recital 71 of the GDPR; we believe that it is possible to talk about a "right to explanation" regulation that can be applied to the data subject in certain situations. In fact, in Google v. Spain decision,[7] which was a landmark case regarding the right to be forgotten, the Court of Justice of the European Union (CJEU) interpreted the access and objection rights of the Directive numbered 95/46 together. In our opinion, the different provisions of the GDPR and Recital 71 which, although not binding, provide a means for us to interpret these provisions more clearly, are of a nature that allows creating a "right to explanation" regulation.

In the field of academic literature, as in general, there are those who hold this view (Malgieri and Comande 2017: 243–265; Goodman and Flaxman 2017: 38; Brkan 2019: 91–121; Mendoza and Bygrave 2017: 77–98; Selbst and Barocas 2018: 1085–1139); and there are also those who are of the opinion that the GDPR gives the right to information in various situations, and does not give a "right to explanation" (Edwards and Veale 2017: 18–84; Wachter et al. 2017: 76–99). In the first draft of the GDPR, Article 22 of the accepted text of the GDPR was regulated under the heading

[5] Directive 95/46/EC of the European Parliament and of the Council of 24 October 1995 on the protection of individuals with regard to the processing of personal data and on the free movement of such data, Official Journal L 281, 23/11/1995 (pp. 0031–0050).

[6] Regulation (EU) 2016/679 of the European Parliament and of the Council of 27 April 2016 on the protection of natural persons with regard to the processing of personal data and on the free movement of such data and repealing Directive 95/46/EC (General Data Protection Regulation-GDPR), Official Journal L 119, May 4, 2016.

[7] Judgment of the Court (Grand Chamber), C-131/12, 13 May 2014, https://eur-lex.europa.eu/legal-content/EN/TXT/?uri=CELEX%3A62012CJ0131 (last visited March 15, 2022).

"Measures based on profiling". In its original form, the article was similar to the provision of Article 15 of the Directive numbered 95/46 and only included profiling based on automated processing. In this first draft, Article 20/4 included informing the data subject of the existence of automatic processing and the envisaged effects of such processing on the data subject. However, this expression in the article was moved to Articles 13 and 14 of the GDPR, and the accepted Article 22 was written in a way to emphasize the right to explain rather than the obligation to inform. Article 22, unlike Article 15 of the Directive numbered 95/46, does not only include profiling. It covers all automated processing and decision-making processes.

As similar to Article 22 of the GDPR, Article 15 of the Directive numbered 95/46 is also included. However, when evaluated together with other regulations, it can be said that the regulation in the Directive numbered 95/46 brings a symbolic regime, whereas the regulation in the GDPR brings much broader, stronger and deeper regulations (Kaminski 2019: 208).

5.2.1 *Brief Overview of Articles 13, 14, 15 of the GDPR*

Article 13 of GDPR

Article 13 of the GDPR is regulated in the Third Chapter of the GDPR titled "Rights of the Data Subject". In this chapter, Article 12 is regulated under the title of "Transparent Information, Communication and Modalities for the Exercise of the Rights of the Data Subject".

Article 13 is regulated under the heading "Information to be provided where personal data are collected from the data subject". In the first paragraph of Article 13, the information that the controller shall, *at the time when personal data are obtained*, provide the data subject, where personal data relating to a data subject are collected from the data subject, is enacted. In the second paragraph of Article 13, apart from the information to be provided at the time when personal data are obtained, it is enacted which "further information" necessary to ensure "fair and transparent processing" shall be provided. The sixth sub-paragraph of this paragraph is as follows:

Article 13/2-f of the GDPR: "the existence of automated decision-making, including profiling, referred to in Article 22(1) and (4) and, at least in those cases, meaningful information about the logic involved, as well as the significance and the envisaged consequences of such processing for the data subject".

It is stated that the phrase "meaningful information" in the first part of this sentence includes a clear explanation of the reasons behind the automatic decision, whereas the phrase "the significance and the envisaged consequences of such processing" relates to the intended and future processing activity (Janssen 2019: 21). According to Recital 60, the principles of fair and transparent processing require the data subject to be informed of the existence and purposes of the processing activity. The data controller must provide the data subject with all additional information necessary to ensure fair and transparent processing, by taking into account the specific circumstances and context in which personal data is processed. In addition, the data subject should be informed of the existence of profiling and the consequences of such profiling. In cases where personal data is collected from the data subject, the data subject should also be informed about whether s/he is obliged to provide the personal data and the consequences of this in case s/he does not provide this data. This information may be provided with standardized icons to give a meaningful overview of the intended processing in a way that is easily visible, understandable and clearly legible. Where symbols are presented electronically, they must be machine-readable.

According to Article 13/3 of GDPR, if the data is processed for a purpose other than that for which the personal data is collected, the data subject must be informed about these purposes and other matters as referred to in Article 13/2, prior to this processing activity.

In addition, an important point here is that in order for this information to be requested, the automatic processing of the information regulated in Article 22/1 does not have to "lead to legal consequences for the data subject" or "significantly affect the data subject".

Article 14 of GDPR

Article 14 of the GDPR is regulated in the Third Chapter of the GDPR titled "Rights of the Data Subject". The title of Article 14 is as follows: "**Information to be provided *where personal data have not been obtained* from the data subject**".

In the first paragraph of Article 14 of the GDPR, the information that the controller shall provide the data subject, where personal data have not been obtained from the data subject, is regulated. In this regulation, the expression "at the time when personal data are obtained" in the first paragraph of Article 13 is not included. Because this article regulates the situation, where personal data is not obtained from the data subject. On the other hand, a time frame is determined in the third paragraph of the article. Accordingly, the Data Controller will provide the information, specified in Articles 14/1 and 14/2, within the following time frame:

a. within "a reasonable period" after obtaining the personal data, but at the latest within one month, having regard to the specific circumstances in which the personal data are processed;
b. if the personal data are to be used for communication with the data subject, at the latest at the time of the first communication to that data subject;
c. if a disclosure to another recipient is envisaged, at the latest when the personal data are first disclosed.

Therefore, it is seen that there is a time restriction in terms of sub-paragraph 13/2-f and sub-paragraph 14-2-g.

In the second paragraph of Article 14 of the GDPR, the information required for "fair and transparent processing" is regulated. The seventh sub-paragraph of this paragraph is regulated exactly the same as the sub-paragraph 13/2-f.

In Article 14/2-f of the GDPR, it is regulated that the controller shall provide the data subject "the information from which source the personal data originate, and if applicable, whether it came from publicly accessible sources". It is possible to deduce from this expression in Article 14 that the data controller collects the personal data of the data subject, but collects it from sources other than the data subject. Therefore, in this case, we think that it is correct to interpret the only use of the phrase "envisaged consequences" together with the phrase "at least" in the same sub-paragraph. The data controller is expected to provide "at least" information about the significance and the envisaged consequences of the processing activity for the data subject. If, as a result of automatic processing tools, there is a result that violates the fundamental rights and freedoms of the person, there should be no doubt that this effect is "significant" for the person and has a concrete effect beyond "envisaged". Therefore, we think that the data controller should make additional explanations in this regard, as there are situations where the expression "at least" can be exceeded.

Article 15 of GDPR

Article 15 of the GDPR, which is in the same chapter as Articles 13 and 14, is enacted under the title "Right of Access by the Data Subject". In Article 15/1, it is enacted that the data subject shall have the right to obtain from the controller confirmation as to whether or not personal data concerning him or her are being processed, and, where that is the case, access to the personal data and the information. Accordingly, in addition to information such as processing purposes, relevant personal data categories, storage period of personal data; the sub-paragraph 15/1-h is regulated just like the sub-paragraphs 13/2-f and 14/2-g.

It is clear that the information to be provided by Articles 13/2-f, 14/2-g and 15-1-h of the GDPR should provide meaningful information about the logic, as well as the existence of automatic decision-making. It is stated that meaningful information should not consist of expressions reflecting the opacity of techniques and algorithms, but should also be understandable (Dalgıç 2020). It is emphasized by the French Data Protection Authority (CNIL) that the phrase "meaningful" is "the capacity to understand the general logic that supports the way the algorithm works". Therefore, instead of using lines of code, it is stated that the processing logic should be explained in easy-to-understand terms by everyone (Dalgıç 2020). In Article 12/1 of the GDPR, which regulates the basic principle of transparency and procedures for information disclosure, the information and communication in Articles 13, 14, 15, 22 and 34 should be "in a concise, transparent, intelligible and easily accessible form, by using a clear and plain language". This statement also reveals that the information about automatic data processing processes in GDPR should be understandable for the data subject. The first and third sentences of Recital 63 of GDPR are also important at this point.

5.2.2 *Examination of Article 22 of GDPR*

Article 22/1 of GDPR

According to Article 22/1 of the GDPR, the data subject shall have the right not to be subject to a decision based solely on automated processing, including profiling, which produces legal effects concerning him or her or similarly significantly affects him or her. Automatic processing is not defined in the GDPR. On the other hand, it is stated by the data protection authorities that the automatic decision-making refers to the process of decision-making by automated means without any human intervention (UK, Commissioner's Office 2022). Today, it is possible to say that such automatic decisions are usually made by machine learning systems, containing artificial intelligence. These decisions can be based on real data as well as digitally generated profiles and inferred data. An example of automated decision-making is the automatic conclusion of a loan application with an online decision, and recruitment processes that are finalized using pre-programmed algorithms and criteria. Automatic decision-making often includes profiling, but it is not mandatory. Profiling is defined in Article 4(4) of the GDPR, as follows:

> 'profiling' means any form of automated processing of personal data consisting of the use of personal data to evaluate certain personal aspects relating to a natural person, in particular to analyze or predict aspects concerning that natural person's performance at work, economic situation, health, personal preferences, interests, reliability, behavior, location or movements.

It can be said that there are four conditions in general for the implementation of Article 22/1 of GDPR:

a. There must be a decision made regarding the person.
b. The decision must be based solely on automated processing.
c. The decision must produce legal effects concerning the person (or)
d. The decision must affect the person significantly.

We will touch on these four issues under two headings:

a. **A decision based solely on automated processing**

The scope of the expression "solely" in the text of the article is discussed here. It is debated whether human involvements, albeit to a small extent, affect the application of this article. In our opinion, the expression "based solely on automated processing" emphasizes that the decision-making process is "automatic". According to the Guide published by the European Data Protection Supervisor, "automatic decision making" is the ability to make decisions with technological means without human involvement (Guidelines 2017).

A passive "human" decision, based solely on algorithmic decision-making, will not prevent falling under GDPR. Whether a decision is "only automatic" will depend primarily on whether human involvement in the decision-making process is technically possible. If human involvement is "impossible", the decision-making process will be considered as "only automatic". Even if there is human involvement, if the human involvement only results in the formal approval of the decision, then we think that it would be correct to characterize the process as "only automatic". Human judgment must be such that it validates the machine-generated decision, so that the decision is not based solely on automated processing (Brkan 2019: 10).

It should be noted that minor human involvements that do not affect the decision-making processes will not prevent the implementation of the article, whereas human involvements that affect the automatic formation of decision-making processes will not be included in the scope of Article 22/1. While the UK Information Commissioner's Office made a statement regarding the implementation area of the article, it stated that the expression "based solely on automated processing" did not include any human involvement (UK, Commissioner's Office 2022). If the decision made by the automated process is examined by the human before the final decision is made and a decision is made by considering it together with other factors, this decision will not be a "decision based solely on automated processing". On the other hand, abuse of this provision by data controllers should also be prevented. If it is said that decision based on automated processing will assist human decision, but the decisions based on automated processing are routinely made in practice without the effect of human participation; then the prohibition in Article 22 of GDPR should be applied. Passive human involvements, based solely on algorithmic decision-making, will not hinder the implementation of the article (Janssen 2019: 19).

Human involvement needs to be meaningful. The Data Protection Impact Assessment (DPIA) application also requires the data controller to reveal the degree of human involvement in the decision-making process and at what stage it occurs.

However, the data controller should not try to go beyond Article 22, by stating that there is "human involvement". For example, if the data controller company routinely applies automatically generated profiles to individuals without any actual influence on the decision, it will only be a decision based solely on automated processing. In order to consider any human involvement sufficient, the data controller should seek that it is "meaningful" rather than a "token gesture" (Guidelines 2017: 10).

A company cannot go beyond the scope of Article 22/1 of the GDPR, by only making human-controlled algorithmic decisions. Human involvement/supervision should be made by a person, "who has the authority and competence to change the decision" (Kaminski 2019: 201; Edwards and Veale 2017: 3).

b. **A decision which produces legal effects concerning him or her or similarly significantly affects him or her**

The phrase "produce legal effects" in the text of the article includes only the decision based solely on automated processing, affecting legal rights such as establishing a legal relationship with others or taking legal actions. A legal effect can be an issue that affects a person's legal status or rights under a contract, as well as legal situations such as contract termination, tax assessment and cessation of citizenship.

At first glance, it seems easier to identify circumstances "producing legal effects" than to determine "significantly effective" decisions. However, due to the phrase "similarly" at the beginning of the "significantly effective" decisions, it would be correct to understand such decisions as having a significant impact equivalent to legal consequences, even if they do not have legal consequences. Even if there is no change in the person's legal rights or obligations or legal status, the decision made as a result of automatic processing can significantly affect the person. The expression "similarly" in the article was not included in Article 15 of the Directive numbered 95/46. This expression necessitates that the "legal effect" be equivalent/similar to legal consequences. In Recital 71 on this issue, "automatic rejection of online loan application, rejection of remote job application" are regarded as examples. In this context, the decisions that significantly affect the conditions, behaviors and choices of the individual, create a long-term or permanent effect on the data subject, cause discrimination, and exclude individuals or groups will be considered as the decisions that have significant effects. The decisions that affect the prevention of access to health services, the refusal of loan requests, and the employment or education status should also be evaluated within the scope of Article 22/1 in general. It is possible to evaluate advertising practices based on profiling in this context, according to the characteristics of the concrete situation. Because in this case, if the data subject does not take into account such targeted advertisements, it will not be possible to say that these decisions affect the data subject significantly (Brkan 2019: 10).

5.2.3 Article 22/2 of GDPR

The circumstances regulated by Article 22/1 of GDPR, in which the data subject's right not to be subject to a decision based solely on automated processing cannot be exercised, are included in Article 22/2 of GDPR. Unless one of the exceptions in Article 22/2 of GDPR applies, the data controller should not undertake the processing based solely on automated activity as specified in Article 22/1.

Accordingly, if a decision based solely on automated processing, including profiling, which produces legal effects concerning the data subject or similarly significantly affects the data subject;

a. is necessary for entering into, or performance of, a contract between the data subject and a data controller;
b. is authorized by Union or Member State law to which the controller is subject and which also lays down suitable measures to safeguard the data subject's rights and freedoms and legitimate interests; or
c. is based on the data subject's explicit consent;

the right of the data subjects not to be subject to the decision is abolished.

a. **if the decision is necessary for entering into, or performance of, a contract between the data subject and a data controller**

Data controllers may wish to use only automated decision-making processes for contractual purposes. Routine human interventions may be impossible due to the sheer volume of data processed. The controller must then be able to prove that such processing is necessary. If there are other effective and less intrusive methods to achieve the same aim, then there will be no "necessity". For example, a company may consider it "necessary" to carry out pre-selection processes by using fully automated tools in order to conclude a contract with the data subject by screening tens of thousands of applications for a vacancy.

b. **if the decision is authorized by Union or Member State**

Automated decision making, including profiling, may become possible, where the Union or Member State laws allow such processing. According to Recital 71, the scope of decision-making processes based on such processing includes monitoring and prevention purposes such as fraud and tax-evasion, carried out under the Regulations, Standards and Recommendations of the EU institutions or national oversight bodies, and the activities carried out to ensure the security and reliability of a service provided by the data controller. In this option, it can be said that the relevant EU regulation or national law should include suitable measures to protect the rights, freedoms and legitimate interests of the data subject.

c. **Explicit Consent**

If the data subject has explicit consent to a decision based solely on automated processing, including profiling; even if the decision produces legal effects concerning him or her or similarly significantly affects him or her, s/he shall not have the right not to be subject to a decision. As a preliminary assumption, Article 22/1 of the GDPR involves high data risks, so individual control over one's own data is seen as an appropriate option.

The consent required for the processing of personal data was regulated in Article 7/1 of the Directive numbered 95/46 as unambiguously and express consent (95/46, Rec. 33). In the GDPR, the issue of consent is regulated in a much more detailed and voluntary manner. In accordance with Article 7/2 of GDPR, if the data subject gives his/her consent with a written declaration, the request for consent must be presented as clearly distinguishable from the other matters, in an intelligible and easily accessible form and by using clear and plain language. The data controller must prove the existence of consent (GDPR, Article 7/1). When assessing whether the consent is given freely, it should be taken into account whether the consent is given in relation to the performance of a contract, including the provision of a service (GDPR, Article 7/4).

In Article 32 of the GDPR, it is regulated that the consent can be given in clear affirmative act establishing a freely given, specific, informed and unambiguous indication, and in writing, electronically or orally. In any case, the data subject must have sufficient knowledge of what s/he is consenting to. It is essential that the data controller adequately inform the data subject about the processing and use of personal data.

5.2.4 *Article 22/3 of GDPR*

In the cases referred to in the sub-paragraphs (a) and (c) of Article 22/2 of the GDPR, in other words, "if the decision is necessary for entering into, or performance of, a contract between the data subject and a data controller" and "if the decision is based on the data subject's explicit consent"; the data subject shall not have the right not to be subject to a decision based solely on automated processing, and the data subject, who is subject to a decision based solely on automated processing activity, will acquire certain rights. In this case, the data controller will be obliged to take suitable measures to protect the fundamental rights and freedoms and the legitimate interests of the data subject.

In the article, the "suitable measures" that the data controller should take are stated as a minimum and not limited number, as follows:

- right to obtain human intervention
- right to express his/her point of view
- right to contest the decision.

The right of the data subject to obtain human intervention should be interpreted as the right of the data subject not to be subject to fully automated processes by entering the human element, instead of a fully automatic decision.

The right to contest the decision obliges adequate information. In case of this possibility, it is not clear enough how the contest will be used and how the contest will be decided. In the event of a contest by the data subject, the data controller will no longer be able to process personal data, unless compelling legitimate grounds for making, exercising or defending legal claims or for processing activities that outweigh the interests, rights and freedoms of the data subject are presented (GDPR, Article 21/1).

When these three rights, which are regulated as a minimum, are considered together with Articles 13/2-f, 14/2-g and 15/1-h of the GDPR, it should be remembered that the data subject also has the following rights:

i. Is the data controller involved in the automated decision-making process or not? The data subject will be able to get information about its existence.
ii. "Meaningful" information about the logic in the processing.
iii. Importance of the processing activity for the data subject.
iv. The envisaged consequences of the processing activity for the data subject.

We think that the ones listed here are at least (minimum) and that the data subject may request other information from the data controller according to the requirements of the situation.

We think that all of these rights give rise to a "right to explanation". In Recital 71 of GDPR, in any case, being subject to a decision based solely on automated processing activity should be subject to the suitable safeguards, including the right to obtain specific information and human intervention assistance for the data subject, to express their own views, to obtain an explanation of the decision reached after such assessment and the right to contest the decision. The Justifications in the EU Regulations are not binding like the articles of the Regulation. However, they are helpful in interpreting the articles of the Regulation. We would like to state that Recital 71 of the GDPR is complementary to the right of the person in Article 22/1 of the GDPR, not to be subject to a decision based solely on automated processing; and that the rights in Article 22/3 given to the data subject, who is subject to such data processing, are a right of explanation beyond the mere right to contest.

5.2.5 Article 22/4 of GDPR

The general rule in Article 22/1 of GDPR, which gives the right not to be subject to a decision based on automated processing, covers all personal data of the person. In Article 22/4, if the "processing of special categories of personal data" is subject to a decision based on automated processing, it is regulated the rules that it will be subject to.

Any kind of data that will make individuals identifiable or determinable is personal data. On the other hand, personal data is generally divided into two as "general personal data" and "special personal data". Special category data refers to personal data that needs more protection due to their sensitive nature, so they are also referred to as "sensitive personal data". According to Article 9 of the GDPR, the processing of personal data revealing racial or ethnic origin, political opinions, religious or philosophical beliefs, or trade union membership, and the processing of genetic data, biometric data for the purpose of uniquely identifying a natural person, data concerning health and data concerning a natural person's sex life or sexual orientation are included within the scope of sensitive personal data.

In order for special category data to be processed in accordance with the law, it must comply with the conditions in Article 6 of the GDPR as well as the special regulations in Article 9. In case that this data is processed based on automated processing, Article 22 will also be applied.

In order for special categories of personal data in Article 22/4 to be subject to an automated processing, the conditions in the sub-paragraphs (a) and (g) of Article 9 must be met. In addition, it is essential to take suitable measures to safeguard the data subject's rights and freedoms and legitimate interests.

Therefore, if the personal data is under "special categories of personal data", the conditions in Article 22/2 alone will not be sufficient, and it is necessary to take suitable measures in order to safeguard the data subject's rights and freedoms and legitimate interests with Articles 9/2-a and 9/2-g. In addition, the principles in Article 6 of the GDPR, which sets out the general principle regarding the processing of personal data, must also be taken into account.

5.3 Conflict with Intellectual Property and Protection of Trade Secrets

It may be possible for data controllers to want to get rid of these obligations by characterizing algorithms as trade secrets or intellectual property. According to Recital 63 of GDPR, the access right of the data subject, regulated in Article 15 of GDPR, should be used without adversely affecting the rights such as trade secrets and intellectual property. However, a balance-based practice is adopted here. It is stated that the data controller cannot refuse to provide all information to the data subject, by claiming these rights.

Trade secrets are mostly on the agenda in algorithmic decision-making systems. However, in recent years, it is observed that algorithms, which provide a technical solution to a technical problem, can be patented to a certain extent. As a general rule in patent law, scientific theories, principles, mathematical methods, business plans and pure mathematical algorithms do not constitute the subject of invention (Turkish Industrial Property Law, Article 82/2-a) (Soysal 2019: 427–453). In addition, if an invention is created as a result of the application of mathematical algorithms, this invention may be entitled to a patent. However, even in this case, there is no problem in terms of algorithmic transparency. Because the inventor has to explain the components of the algorithm and the way it works in the patent application. The problem arises in trade secret protection. As the algorithms are covered by trade secret protection, it prevents algorithmic transparency. In this regard, in Article 5 of the EU Directive No. 2016/943 on the Protection of Trade Secrets,[8] in which the exceptions are regulated; there is a provision that trade secret protection can be disabled "for the purpose of protecting a legitimate interest recognized by the Union or national law". The right to make a statement regarding an automatic decision that leads to algorithmic discrimination can be considered as "protecting legitimate interest" in this context. However, even if it is not within this scope, it is regulated that trade secret-like protections will not completely prevent the data subject from being given information (GDPR, Rec. 63).

Another issue that can hinder algorithmic transparency may be state secrets or information kept confidential for public reasons. In some cases, algorithmic transparency may not be achieved for public reasons. Even in these cases, we think that the data subject has the right to be reasonably informed about the decision made by the algorithm about him, without touching the essence of the secret.

[8] Directive (EU) 2016/943 of the European Parliament and of the Council of 8 June 2016 on the protection of undisclosed know-how and business information (trade secrets) against their unlawful acquisition, use and disclosure, Official Journal of the European Union, L 157/1, June 15, 2016.

5.4 Conclusion and Evaluation

It is not explicitly mentioned about the right to explanation anywhere in the GDPR, other than Recital 71. **However, we think that it is possible to talk about a "right to explanation", when the whole GDPR is evaluated together with Articles 13(2)(f), 14(2)(g), 15(1)(h) and especially Article 22. In the new opacity created by algorithms, the data subjects have the right to receive information on three bases:**

i. The data subjects, involved in automated processing but not subject to an automated decision as they meet the definition in Article 22/1 of the GDPR, will be able to receive information about the existence of automated decision-making based on Articles 13/2-f, 14/2-g, 15/1-h of GDPR. Recitals 60 and 63 also help us with this interpretation.
ii. The data subjects, who meet the definition in Article 22/1 of GDPR and are subject to an automated decision, will be able to receive meaningful information about the logic involved, as well as the significance and the envisaged consequences of data processing, except for the existence of automated decision-making based on Articles 13/2-f, 14/2-g and 15/1-h of the GDPR.
iii. The data subjects, who meet the definition in Article 22/1 of GDPR and are subject to an automated decision, will have the right to receive special information and explanations based on Article 22/3 of GDPR. In this regard, the data subject shall have the right to request to implement suitable measures to safeguard his/her rights and freedoms and legitimate interests, the rights to obtain human intervention, to express his or her point of view and to contest the decision.

First of all, Article 22 applies to automated decision-making processes, where there is no human intervention. Although technological possibilities allow in the face of increasing criticism in recent years, there are not many algorithms that make decisions as a result of fully automatic processes without human intervention. In fact, the tendency to benefit from such algorithms as an assistant is increasing, by considering the situations that usually involve the risk of discrimination. In other words, the human makes the final decision. However, it should be kept in mind that this is the case in theory, and in practice, it is decided on the basis of the determinations of such automatic systems. It is clear, however, that the wording of Article 22 creates an ambiguity for situations involving human intervention (Edwards and Veale 2017: 44).

It is also discussed that Article 22 requires a "decision" that "producing legal effects" or "affecting significantly" (Edwards and Veale 2017: 46). First of all, in such automated systems, it is not clear in which situations a decision should be made. It is asked whether the outputs of machine learning systems containing artificial intelligence should be accepted as a decision, or if these outputs do not constitute a final decision, whether they will still be accepted as a decision only if they are an auxiliary tool, for example, to the police or judge. When we interpret the process technically and accept every result produced by automatic systems as a decision, the criteria of "producing legal effects" or "affecting significantly" may not be met. Therefore, in this process, it would be more accurate to make a determination related with the person based on the criteria of "producing legal effects" and "affecting significantly". However, it is a fact that both of these conditions restrict the implementation of this article.

The right to explanation in GDPR covers only data subject individuals. It does not include companies. Sometimes companies also provide input on products or services. It is stated that the companies should also have the right to receive explanations about how their inputs affect the decisions, made as a result of automated processing (Janssen 2019: 28). Individuals, who are apart from the data subject but whose data are collected as samples in large-scale decision-making systems, do not have the right to receive explanations. These persons are not directly subject to a decision based solely on automated processing with similar significant effects. In addition, they are included in the automatic data processing. However, they do not benefit from the right to explanation. These persons have the right to receive information about the existence of automated processing in accordance with Articles 13/2-f, 14/2-g and 15/1-h of the GDPR. They do not have the right to receive meaningful information about the logic involved, as well as the significance and the envisaged consequences. They also do not have a right to receive information based on Article 22/3 of the GDPR (Janssen 2019: 28–29). It is not possible for the general public to benefit from the right to explanation. Whereas, the right to explanation for public in general may strengthen the concept of informed consent regarding automated decision-making. Therefore, it can be said that the right to explanation does not increase transparency for all parties involved in the algorithmic decision-making process.

Due to the expression in Articles 13 and 14 of the GDPR, it is stated that these articles require information to be given about the functions of the automatic system as ex ante, whereas Articles 15 and 22/3 of the GDPR require an explanation as ex post about a certain decision together with the functions of the automatic system (Janssen 2019: 30). Transparency requires both ex ante and ex post explanations. Algorithms are somewhat opaque by nature, which means that they are not transparent. An ex ante explanation often provides incomplete information about the ways and purposes for which the algorithmic decision-making system uses personal data. On the other hand, "ex post" explanations make it possible to give more descriptive information to the data subject. Ex post explanations describe the actual factors that the algorithm used, instead of the default factors, to make a decision. In addition, the explanations can be given that enable the data subject to contest to the decision. However, as we stated in the text of the article, we believe that despite this uncertainty, it is possible for the data subject to request broader explanations from the data controller.

Despite these shortcomings, it is possible to accept the regulations on the right to explanation in GDPR as a powerful instrument that can be used by the individuals regarding algorithmic discrimination. Our comments on the regulations, expressed in the text of the article, also support this idea.

It is also of great importance for us that National Equality Institutions and Data Protection Authorities have adequate investigative and enforcement powers in preventing discrimination based on artificial intelligence. Because the inherent opacity and technical difficulties in algorithms will take much more time to be resolved by the courts. It is important that specialized institutions in this field have the necessary equipment with regard to this issue. In this context, it seems necessary to make amendments, similar to the right to explanation, in Turkish Personal Data Law.

Indeed, it is not possible to say that the regulation on the right to explanation in GDPR alone will eliminate algorithmic discrimination or that it includes all remedies. Just as there was an industrial revolution, a single law could not solve the problems. In the new age transformed by artificial intelligence, we will not be able to find solutions to problems with a single law. It seems inevitable to analyze different risks and make additional amendments in a way to include new types of discrimination. However, we believe that it is possible to use GDPR regulations as an effective instrument against algorithmic discrimination in this process.

We would like to complete our article with a short story told by Eduardo Galeano (Galeano 2004: 85)[9]:

> A lumberjack, by going out for work, realizes that his ax is missing. He keeps his eyes peeled to his neighbor and concludes that his mimics, delivery and looks show perfectly his theft. Some days after, he finds his ax in the forest where he did drop. Again, when he observes his neighbor, he thinks that he doesn't resemble to a thief a bit, with neither his mimics nor delivery or looks.

Therefore, we should not forget that our bias always exists within us.

References

Aneesh A (2002) Technologically coded authority: the post-industrial decline in bureaucratic hierarchies. In: Conference Paper. http://web.stanford.edu/class/sts175/NewFiles/Algocratic%20Governance.pdf (last visited March 1, 2022)

Borgesius FJZ (2020) Strengthening legal protection against discrimination by algorithms and artificial intelligence. Int J Human Rights 24(10):1572–1593. https://www.tandfonline.com/doi/pdf/10.1080/ (last visited March 15, 2022)

Brkan M (2019) Do algorithms rule the world? Algorithmic decision-making in the framework of the GDPR and beyond. Int J Law Inf Technol 27(2):91–121

Dalgıç Ö (2020) Algorithms meet transparency: why there is a GDPR right to explanation? April 8, 2020. https://turkishlawblog.com/read/article/221/algorithms-meet-transparency-why-there-is-a-gdpr-right-to-explanationg (last visited March 15, 2022)

Danesi M (2021) Pythagoras' legacy: mathematics in ten great ideas, Ketebe.

[9] "Indicios No se sabe si ocurrió hace siglos, o hace un rato, o nunca.

A la hora de ir a trabajar, un leñador descubrió que le faltaba el hacha. Observó a su vecino y comprobó que tenía el aspecto típico de un ladrón de hachas: la mirada, los gestos, la manera de hablar...

Unos días después, el leñador encontró su hacha, que estaba caída por ahí.

Y cuando volvió a observar a su vecino, comprobó que no se parecía para nada a un ladrón de hachas, ni en la mirada, ni en los gestos, ni en la manera de hablar".

Edwards L, Veale M (2017) Slave to the algorithm, why a 'right to an explanation' is probably not the remedy you are looking for? Duke Law Technol Rev 16
Engle K (2016) How the invisible hand of technology can help us make better decisions, February 24, 2016. https://socialmediaweek.org/blog/2016/02/invisible-hand-technology-can-help-us-make-better-decisions/ (last visited March 1, 2022)
Galeano E (2004) Bocas Del Tiempo, Catalogos
Goodman B, Flaxman S (2017) European Union regulations on algorithmic decision-making and a right to explanation. In: International conference on machine learning, workshop on human interpretability in machine learning, vol 3. AI Magazine, p 38
Guidelines (2017) Guidelines on automated individual decision-making and profiling for the purposes of regulation 2016/679, Article 29, Data Protection Working Party, October 3, 2017. https://ec.europa.eu/newsroom/document.cfm?doc_id=47742 (last visited 15 March, 2022)
Janssen JHN (2019) The right to explanation: means for 'White-Boxing the Black-Box', Tilburg University, LLM Law and Technology, January 2019. http://arno.uvt.nl (last visited March 15, 2022)
Kaminski M (2019) The right to explanation, explained. Berkeley Technol Law J 34:189–218. https://papers.ssrn.com/ (last visited March 14, 2022)
Mac Sithigh D, Siems M (2019) The Chinese social credit system: a model for other countries, European University Institute, Department of Law, January 2019. https://cadmus.eui.eu/bitstream/handle/1814/60424/LAW_2019_01.pdf (last visited March 15, 2022)
Malgieri G, Comande G (2017) Why a right to legibility of automated decision-making exists in the general data protection regulation. Int Data Privacy Law 7(4):243–265. https://doi.org/10.1093/idpl/ipx019 (last visited MArch 15, 2022)
Mendoza I, Bygrave LA (2017) The right not to be subject to automated decisions based on profiling. In: EU internet law, regulation and enforcement, pp 77–98
Minowitz P (2004) Adam Smith's invisible hands. Econ J Watch 1(3):381–412. https://econjwatch.org/File+download/268/ejw_com_dec04_minowitz.pdf (last visited, March 15, 2022)
O'neil C (2016) Weapons of math destruction: how big data increases inequality and threatens democracy. Crown
Pasquale F (2020) New laws of robotics, defending human expertise in the age of AI. Belknap
Pasquale F (2015) The black box society: the secret algorithms that control money and information. Harvard University Press
Rainie L, Anderson J (2017) Code-dependent: pros and cons of the algorithm age, February 8, 2017. https://www.pewresearch.org/internet/2017/02/08/code-dependent-pros-and-cons-of-the-algorithm-age/ (last visited March 1, 2022)
Selbst AD, Barocas S (2018) The intuitive appeal of explainable machines. Fordham Law Rev 87(3):1085–1139. https://ir.lawnet.fordham.edu (last visited March 15, 2022)
Smith A (2022) The wealth of nations, Book 1, Chapter 2. https://geolib.com/smith.adam/won1-02.html (last visited March 1, 2022)
Soysal T (2019) Agricultural biotech patent law. Adalet Publications
Soysal T (2020) Industry 4.0 and human rights: the transformative effects of new emerging technologies on human rights. In: Ankara Bar Association's 11th international law congress, January 9–12, 2020, Congress Book, vol 1, pp 245–325. http://www.ankarabarosu.org.tr/Siteler/2012yayin/2011sonrasikitap/2020-hukuk-kurultayi-1-cilt.pdf (last visited March 15, 2022)
Tay (2016) Microsoft's AI chatbot, gets a crash course in racism from Twitter, March 24, 2016. https://www.theguardian.com/technology/2016/mar/24/tay-microsofts-ai-chatbot-gets-a-crash-course-in-racism-from-twitter (last visited March 1, 2022)
The Sterling 'flash event' of 7 October 2016, Markets Committee. https://www.bis.org/publ/mktc09.pdf (last visited, March 1, 2022)

UK, Information Commissioner's Office (2022) What does the UK GDPR say about automated decision-making and profiling. https://ico.org.uk/for-organisations/guide-to-data-protection/guide-to-the-general-data-protection-regulation-gdpr/automated-decision-making-and-profiling/what-does-the-uk-gdpr-say-about-automated-decision-making-and-profiling/ (last visited February 11, 2022)

Wachter S, Mittelstadt B, Floridi L (2017) Why a right to explanation of automated decision-making does not exist in the general data protection regulation. Int Data Privacy Law J 7(2):76–99. https://doi.org/10.1093/idpl/ipx005 (last visited March 12, 2022)

Tamer Soysal He graduated from Ankara University Faculty of Law in 1999. He received his MBA from Erciyes University, Institute of Social Sciences in 2004 with the thesis titled "Protection of Internet Domain Names"; He completed his Ph.D. in Private Law at Selcuk University, Social Sciences Institute in 2019 with the thesis titled "Applications of Biotechnology in Agriculture and the Patentability of Such Inventions". In addition, in 2006, he completed the Sports Law Program organized by Kadir Has University, Sports Studies Center with his thesis titled 'Betting Games in Sports'. He has many published articles in the field of informatics and intellectual property law. He worked as a Public Prosecutor for about 10 years and started to work as an Investigation Judge in the Human Rights Department of the Ministry of Justice in 2012. Currently, he is the Head of EU Project Implementation Department at the Ministry of Justice.

Part III
Evaluation of Artificial Intelligence Applications in Terms of Criminal Law

Chapter 6
Sufficiency of Struggling with the Current Criminal Law Rules on the Use of Artificial Intelligence in Crime

Olgun Değirmenci

Abstract Every new technology affects crime, which is a social phenomenon. This interaction is in the form of either the emergence of new forms of crime or the facilitation of committing the crime. Based on the definition of intelligence as the ability to adapt to changes, artificial intelligence is defined as "the ability to perceive a complex situation and make rational decisions accordingly". Based on this definition, in cases where the decisions taken constitute a crime, it is necessary to determine the responsibility in terms of criminal law. The criminal responsibility of artificial intelligence may immediately come to mind. However, holding artificial intelligence, which does not form a legal personality, responsible in terms of criminal law is a controversial situation. Secondly, the responsibility of the software developer who created the artificial intelligence algorithm can be discussed here as well. And yet, in this second case, the willful or negligent responsibility of the software developer should be examined separately. In terms of the negligent responsibility of the software developer who created the artificial intelligence algorithm, the issue of whether artificial intelligence can be used in committing a crime is predictable should be addressed. In this paper, it will be examined whether the existing regulations will be sufficient to determine the responsibility in terms of criminal law where the artificial intelligence algorithm is used in the commission of a crime.

Keywords Artificial intelligence · Crime · Criminal law · Turkish criminal code · Criminal responsibility

JEL Codes K14 · K41

O. Değirmenci (✉)
Faculty of Law, TOBB Economy and Technology University, Ankara, Turkey
e-mail: odegirmenci@etu.edu.tr

M. Kılıç and S. Bozkuş Kahyaoğlu (eds.), *Algorithmic Discrimination and Ethical Perspective of Artificial Intelligence*, Accounting, Finance, Sustainability, Governance & Fraud: Theory and Application, https://doi.org/10.1007/978-981-99-6327-0_6

6.1 Introduction

After the second half of the twentieth century, when computers, or with a more inclusive term, information systems, began to play an important role in social life, a role change began to take place. The change in question has taken place in the form of information systems moving from being a human tool to functioning themselves instead of humans. Now, it is mentioned that software with human features can make decisions on behalf of humans and implement them like humans.

The traces of the ideas about the contribution of machines with human abilities to human life by doing work instead of humans can be traced back to ancient times (Dyson 1997: 7). For example, it is possible to follow these traces from the self-moving chairs called "tripod" in Homer's Iliad, to Aristotle's longing and Hobbes' idea of building an "artificial animal" (Nilsson 2018: 19).

It may be possible to use the aforementioned software called artificial intelligence today, which can make decisions instead of humans in committing the crime, which is a social phenomenon. Going one step further, the software in question is also likely to cause social harm, as they can make decisions instead of people. In this case, the subject of our communiqué is whether the means of criminal law are sufficient for the reparation of the damage.

6.2 The Concept of Artificial Intelligence

It is stated that artificial intelligence was first introduced as a concept in 1955 by the American computer scientist John McCarthy. The concept in question basically refers to the work that requires a certain intelligence to be done by machines actually done by humans (Bak 2018: 212).

Intelligence is defined as problem-solving ability; it includes perception, understanding, comprehension, reasoning, learning, and problem-solving processes as a necessity. From this point of view, artificial intelligence is "*... machines that can think like humans and make decisions on their own, as well as have the ability to do the work that people focus on and solve the problems they are trying to solve*"[1] (Değirmenci 2021: 74), "*thinking, decision-making, problem-solving and learning associated with humans*" (Değirmenci 2021: 74). "*Automation of activities such as "automation*" (Bellman 2015: 6), "*computational studies that enable perception, rationality and movement*" (Winston 1992). These definitions are also used by "silicon ethics" advocates, who argue that the scope of entities that can be recognized as subjects by law should be expanded to include intelligent computer programs.

By analogy with the human brain, proponents of the "silicone ethics" accept three widely expressed basic propositions. First of all, the human brain is an information processing tool. Secondly, the information processing of the brain can be learned.

---

[1] The definition in question is the definition made by John McCarthy, who used the concept for the first time, at the Dartmouth Conference in 1955. Refer to details: (Değirmenci 2020, 2021).

Finally, every definable process can be recreated on the computer (Michalczak 2017: 94, 94).

Artificial intelligence has different fields of study. In cases where the learning processes of living things are not sufficiently known, empirical systems or artificial neural networks that work by calculating the correlation between inputs and outputs are used. In cases where there is a system but we do not know about its past outputs, we use the behavior model estimation; Heuristic algorithms that produce solutions based on intuition can be counted among them (Esin 2019). However, the issue that is more important in terms of our paper is the types of artificial intelligence. Artificial intelligence is divided into two sub-types: narrow (weak) and broad (strong) artificial intelligence. In a narrow sense, artificial intelligence is hardware and software that is not conscious of itself and acts according to the codes written by its programmers. Strong artificial intelligence, on the other hand, is defined as machines that can perform all the mental activities that humans can do, and in a sense, think (Bacaksız and Sümer 2021: 25; Taşdemir et al. 2020: 798; Miailhe and Hodes 2017; Tegmark 2019: 44). Here, the issue of performing the learning process with the data processed by strong artificial intelligence and updating its algorithm accordingly should be handled with utmost importance. In this case, we are faced with an entity that designs the software itself and dominates the cultural evolution process. This is comparable to the life 2.0 that Tegmark refers to.

6.3 The Question of Whether Artificial Intelligence Can Be a Legal Person

Artificial intelligence can be used in crime in two ways. First of all, it may be possible to use artificial intelligence as a tool in committing a crime. In this case, there is actually not much problem to be solved. As a matter of fact, in this possibility, artificial intelligence is used as a tool by humans in committing the crime.

As for the second possibility, it is necessary to examine the problem of whether artificial intelligence can be the perpetrator of a crime. This possibility is naturally practicable in terms of advanced or strong artificial intelligence. In examining this possibility, first of all, it will be examined whether artificial intelligence can have a personality and then whether it can be a perpetrator.

As a legal term, person is derived from the Latin word "*persona*". The word in question means "to wear a mask" and is not equivalent to the word "*human/homo*" (Brozé 2017: 4). Personality is the legal status of a person and is independent of other statuses. Thus, a person may have more than one personality (unus homo sustinet plures personas).

The "person" is used in the sense of being entitled to rights (Uzun 2016: 13). In all legal orders, it is seen that human beings are accepted as a person. However, it is known that some legal systems also give personality to non-human beings. In general, in order for an entity to be seen as an entity equivalent to a human being, it must have the following three characteristics; the ability to (1) interact with the environment, think and communicate in a complex way, (2) be conscious of one's own existence in order to achieve the determined purpose in life, and (3) live in a society based on mutual interest with other people (Hubbard 2022; Kılıçarslan 2019: 373; Bacaksız and Sümer 2021: 136).

When strong artificial intelligence is evaluated in terms of these three features, it is seen that it can communicate with humans and other devices, think intricately, try to reach its goal with the awareness of its own existence, and coexist with other systems based on mutual interest. The fact that necessary but not sufficient criteria for personality are provided does not lead to the conclusion that artificial intelligence will have personality in legal systems. First of all, personality is a result of the relevant legal order, and the outcome will be a result of the political decisions of the legislator. As a result of these decisions, artificial intelligence may have rights in legal systems. For example, artificial intelligence that creates a musical work, writes a book or paints, and realizes an original design, may also have the intellectual and industrial rights of the relevant work, provided that it is authorized to have rights in the relevant legal system.

6.4 Opinions on Whether Artificial Intelligence Can Have Legal Personhood

6.4.1 Opinions that Support Artificial Intelligence Can Have Legal Personhood

There are opinions that artificial intelligence should have a legal personhood. The opinions in question can be examined under four different headings: artificial intelligence is a legal entity/personhood, an artificial proxy/representative, an electronic personality, and a non-human person.

The legal personality/personhood view has emerged from cases such as being able to get into debt and finding capital in order to compensate for the damages, especially as a result of the decisions taken by artificial intelligence. Similar discussions were made for trading companies in the early days when they were first established, and capital structures emerged. The increasing influence of commercial companies, especially in Europe, necessitated the making of new legal regulations and in this context, the Council of Europe has issued its Recommendation R 88 (18). Following this decision, some countries have changed their domestic laws and started to regulate the criminal responsibilities of legal persons. Regulations accepting the criminal liability of legal persons were made in the French Criminal Code in 1994, in the Belgian Criminal Code in 1999 and in Denmark in 2002 (Centel 2016; Özen 2003).

The artificial proxy/representative view is built on the view that there is actually a proxy relationship between artificial intelligence and human beings, arguing that artificial intelligence is a proxy representing human beings. In this case, the principal may be held responsible for some of the actions of the proxy (Bacaksız and Sümer 2021: 151).

The electronic personality view has been included in the European Parliament's Recommendation dated 16 February 2017 and numbered 2015/2103 (Nowik 2021: 5). In the aforementioned Decision, an electronic personality model has been proposed to eliminate the damages caused by robots that make autonomous decisions and interact independently with third parties. The said proposal is no longer on the agenda of the European Union (Luzan 2020: 6; Negri 2021: 6).

The view of exceptional citizenship is one of the views put forward in this field in terms of artificial intelligence. Based on some news that were also reflected in the press, it is also stated that the decisions regarding the granting of citizenship or residence permit do not have a legal basis. In particular, the granting of citizenship to the robot Sophia by Saudi Arabia or the granting of residence permit by Japan to the chatbot Shibuya Mirai in 2017 are given as examples (Atabekov and Yastrebov 2018: 776; Bacaksız and Sümer 2021: 154; Jaynes 2020: 343).

6.4.2 *Opinions that Support Artificial Intelligence Cannot Have Legal Personhood*

Opinions accepting that artificial intelligence should not have legal personality differ in terms of which model should be applied to artificial intelligence. In this regard, "the view of property", which argues that it has the status of goods, based on the fact that it is an object for artificial intelligence; "the view of slavery status" based on the status of slaves in Roman law; "the view of limited incompetence" regarding the imposition of limitations on legal capacity will be discussed.

The property view is based on the fact that the artificial intelligence entity is a property/goods. Therefore, whether artificial intelligence appears in the form of a robot with a material asset or only in the form of software, it will be the property of a natural person or legal person. Of course, at this point, it is necessary to point out a distinction in terms of Turkish law. If the artificial intelligence consists of software only, the rules of intellectual property law will be applied, not the rules of property law (Bacaksız and Sümer 2021: 154).

The slavery view analyzes the issue based on the status of the slave in Roman law. In Roman law, the slave was not the subject but the object of the law. He could not own property, nor could he have family rights. Provided that he had the capacity to act, he could take legal action, but the slave's action was considered valid only if it was for the benefit of his master. From this point of view, artificial intelligence beings will be able to act, think, and continue to exist as human servants, just like slaves, and be subject to commercial transactions (Jaynes 2020: 343–346).

The opinion of limited competence, on the other hand, is the TCvC (Turkish Civil Code) art. It is based on 16/2. According to this opinion, in a case where artificial intelligence has the power to distinguish, an area of responsibility can be created by creating a small or limited class (Taşdemir et al. 2020: 806).

6.4.3 *The Relationship of Artificial Intelligence and Crime*

As a social phenomenon and a type of deviant behavior that is sanctioned by law, crime and the phenomenon of artificial intelligence may be related. This relationship may be in the form of the use of artificial intelligence by a natural person in the commission of the crime, in other words, the use of artificial intelligence as a tool. As a second possibility, artificial intelligence, which processes, learns, judges, and makes decisions on behalf of human beings, causes a crime to be committed.

In 1981, a 37-year-old Japanese worker working in a motorcycle factory was killed by an artificial intelligence robot working next to him. The robot erroneously identified the worker as a threat to its task, and the most effective way to get rid of this threat was to operate its hydraulic arms, causing the worker to die by hitting the working machine (Hallevy 2013: xv). As current examples, compilation of personal data over many algorithm information systems using artificial intelligence, violating the privacy of private life or using intellectual and artistic works in a way that violates the rights of the author (Hallevy 2021: 222).

Similarly, the story of the Twitter character Tay of the Microsoft company, which was reflected in the press in 2016, showed that artificial intelligence can actually commit some crimes. Tay communicated with users on Twitter but repeated what it had learned from the accounts it communicated with and made hate speech (Say 2018: 151). In this case, it becomes important who will be responsible for the crimes committed, in other words, who will be the perpetrator in terms of criminal law.

6.4.3.1 Using Artificial Intelligence as a Crime Tool

In this possibility, artificial intelligence was used as a tool in a crime committed. Although it is mentioned that it is used as a tool in the crime, artificial intelligence may have been used deliberately or negligently in the crime committed. Therefore, we will conduct a dual analysis under this heading.

In this context, the first thing that comes to mind is that since we define artificial intelligence as created intelligence, it can be accepted that the person or persons who created artificial intelligence, namely, who code and create the algorithm in question, will be responsible for this situation. In other words, if it predicts that the artificial intelligence will commit the crime in question and creates the relevant code accordingly, it will act deliberately.

Here, if artificial intelligence is created for the purpose of committing a crime, punishment is possible even if the intended crime has not yet started enforcement actions. As a matter of fact, the Cybercrime Convention prepared by the Council of

Europe,[2] under the title of "Misuse of Devices" in its sixth article, production, sale, procurement for use, import, distribution or otherwise making available of devices or software for the committing of cyber crimes regulated in the Convention are penalized.

After the Cybercrime Convention was transposed into domestic law, Article 245a was added to the Turkish Penal Code, especially within the scope of harmonization of the Convention and domestic law (Korkmaz 2018). According to the aforementioned article, making and creating of a device, computer program, password, or other security code in order to commit crimes in the Tenth Chapter titled Crimes in the Field of Informatics, and other crimes that can be committed by using information systems as a tool, the act of manufacturing, importing, consigning, transporting, storing, accepting, selling, offering for sale, purchasing, giving away or keeping are penalized. The article was written extensively, and computer programs were also clearly included in the article.

In this context, artificial intelligence, in case when it commits one of the crimes regulated in TPC Articles 243, 244, 245 or tries to influence the programs that protect the computer programs regulated in the Law on Intellectual and Artistic Works No. 5846 (Değirmenci 2020), the deliberate creation, commercialization and possession of artificial intelligence software are penalized according to the penal norm regulated under TCK Article 245a. As an intermediate result, the creation of an artificial intelligence algorithm for the commission of a crime, the subject of trade or possession of the said algorithm, together with the existence of other conditions in 245a, it will constitute a crime.

In the meantime, we should point out that there are opinions in both national and international law that artificial intelligence, which is in the nature of malicious software created for the commission of a cybercrime, should be accepted as a weapon (Downing 2005: 733).

If artificial intelligence, a software not created to commit a crime, is used to commit a crime, it will be used in a deliberate crime. In this case, the person will be responsible in terms of criminal law by using artificial intelligence as a tool for crime.

As a second possibility, the artificial intelligence was not created to commit crimes at first, but the algorithm, which makes machine learning by processing big data, commits some crimes after a certain period of time. In an example that was also reflected in the press, the artificial intelligence algorithm Tay, which analyzes the tweets sent on Twitter, learned a racist discourse as a result of the analysis and finally, made racist statements toward some people through an account it opened. In this case, the programmer or user may be liable for negligence. If the crime committed is a crime that can be committed negligently, and if the *"limits of obligations of care and attention"* in the creation of artificial intelligence are exceeded, the programmer or the user will be liable for the negligent crime.

[2] The Convention was accepted by the Grand National Assembly of Turkey with the Law No. 6533 dated 22.4.2014 and entered into force after being published in the Official Gazette dated 2.5.2014 and numbered 28988.

Here, we should point out that although the user does not violate the "*obligation of attention and care*" while creating the artificial intelligence, it will not be possible for the user or the programmer to be penalized if the artificial intelligence acts with unforeseen outputs due to the fact that it is an intuitive algorithm from the data it processes outside of foresight. For example, the situation where a heuristic algorithm that detects attacks on a website and activates security software considers continuous service requests from an IP over the Internet as an attack, accessing the relevant web page within the scope of legitimate defense and rendering the web page unusable can be considered in this context. At this point, in order to be able to talk about negligence liability, it is necessary to determine the limits of the "attention and care liability". In our opinion, this limit can be derived from the ethical principles established in the studies in this field and will be developed over time. Therefore, the creation of ethical codes in terms of artificial intelligence studies will be important to determine the criminal responsibilities of those who created the algorithm.

6.4.3.2 The Problem of Whether Artificial Intelligence Can Be a Criminal Perpetrator

The question of whether artificial intelligence can be the perpetrator of a crime, in other words, whether it can knowingly and willingly commit an act defined in the criminal codes will be determined according to the regulations in national laws. However, being the perpetrator of a crime requires that he/she acts consciously and with free will.

When we define free will, in its most general form, as the ability to make independent choices and decisions, the availability of options for free will and the ability to choose between options outside of the predetermined rules are necessary. Although it is early to talk about the future, we can say that today the majority of artificial intelligence systems do not have free will, since they are algorithms created to achieve a predetermined goal (Doğan 2021: 806). However, in the future, in the event that the stratification of deep learning occurs and deep learning increases, it will be possible to say that artificial intelligence has free will, in cases where it is possible to choose between possibilities in a non-deterministic way.

Even if the existence of the free will of artificial intelligence is accepted, it should be examined whether the relevant legal system attributes a legal personality to artificial intelligence. In the absence of legal personality, it will not be possible to say that artificial intelligence can be the perpetrator of a crime.

In the third step, although a legal personality is recognized for the artificial intelligence, it should be clearly determined whether the relevant legal personality can be the perpetrator in terms of criminal law. As a matter of fact, in many legal orders, even if legal entities are accepted as legal entities, it may be regulated that legal entities cannot be perpetrators in terms of criminal law.

Artificial intelligence, which is a software, will be able to perform the action that will cause a change in the outside world. In this respect, it can be stated that the movement sub-element, which is an element of crime, can be realized by artificial intelligence, considering that it will not only originate from human beings. For example, in SCADA systems, artificial intelligence that controls the physical system can cause changes in the outside world. In our opinion, the problem of whether artificial intelligence can act with intent or negligence is directly related to the problem of free will. It can be said that the artificial intelligence, which knows in what way the command it has given will produce results and gives the command, accordingly, realizes the result knowingly and willingly.

Although it is stated that artificial intelligence can be the perpetrator despite all the coercion, the sanction to be applied must be appropriate to its structure. As a matter of fact, improvement, which is the purpose of punishment, will not be a valid goal in terms of artificial intelligence. In terms of artificial intelligence, security measures and especially the protection of society will be a prominent goal.

Artificial intelligence may be subject to its own, regulated security measures. For example, artificial intelligence may be recognized as a personality, and fines may be imposed if it has the right. In addition, the learning of artificial intelligence, which makes machine learning by processing big data and commits crimes according to what it learns, can be deleted by digital confiscation, and hence remove the data causing crime. Again, the prison sentence imposed on real persons can also be applied in the form of restricting the freedom of an artificial intelligence entity by maintaining an autonomous activity under the control of a programmer. In addition, it may be possible to apply the death penalty, as in real people, by erasing artificial intelligence completely, irreversibly. In particular, *Hallevy* states that the permanent rendering of an artificial intelligence entity inoperable will be a sanction similar to the death penalty in natural people (Hallevy 2015).

6.5 Conclusion and Evaluation in Terms of Turkish Law

Artificial intelligence, as a software that can make decisions, solve problems, perceive, comprehend, and reason instead of humans, has started a change that forces legal systems. First of all, it has been discussed whether the entity in question can be a legal personality or not. Although there are different approaches in comparative law, apart from a few examples, there is no example that gives personality to artificial intelligence. The fact that artificial intelligence is not recognized as a personality does not mean to prevent possible problems. As a matter of fact, if artificial intelligence causes social or individual harm, there is a need for rules that determine who will be responsible for the said damage. The aforementioned rules may be applied with an unexceptionable/ordinary approach, according to the characteristics of the existing rules, or new regulations may be needed within the scope of the exceptional approach.

In terms of Turkish law, matters related to personality are regulated in the Turkish Civil Code. In accordance with Article 8 of the TCvC, every human being has been given the capacity to be entitled and it has been stated that he/she is a person. In addition, the Turkish Civil Code regulates legal entities. In addition, since there is no special regulation in terms of artificial intelligence, it cannot be said that artificial intelligence is recognized in Turkish law.

Article 38/7 of the Constitution regulates the principle of the individuality of criminal responsibility. Similarly, in accordance with Article 20 of the Turkish Penal Code, the personality of criminal responsibility has been adopted. When the concept of person in here is connected with the TCvC, it can be concluded at first that natural persons and legal persons have criminal responsibility in Turkish criminal law. However, our country did not accept that legal persons should be perpetrators in criminal law, as in the examples of Germany, Italy, and Spain. The most important reason for this is the need for a legal person's action that will bring about a change in the outside world to be carried out by a natural person at all times. As a legal fiction, a legal entity cannot bring about a change in the outside world by itself. From this point of view, it cannot be a perpetrator. However, in Turkish criminal law, although legal persons cannot be perpetrators, it is possible to apply some security measures in case a crime is committed for the benefit of the legal person. These are confiscation and cancellation of operating license.

Therefore, it is not possible for artificial intelligence to be a perpetrator in Turkish law. In this case, it may be possible to use artificial intelligence as a tool in committing a crime. In other words, the real person who uses artificial intelligence as a tool in a crime will be the perpetrator.

In case of the artificial intelligence is a tool for committing a crime against the will of the person who created the artificial intelligence, the responsibility of the person who created the algorithm can be determined by negligence. In this case, if the person who created the algorithm creates an erroneous code in violation of the objective duty of care or cannot foresee the result of a code, negligent liability may come to the fore. For example, in the creation of an artificial intelligence algorithm that adjusts the red light-green light status according to the vehicle density in traffic, in the event that a faulty code causes the green light to turn on for both directions at the same time, even for a short time, and as a result, a traffic accident resulting injuries occurs, the person who created the algorithm violates the duty of care. Evaluation will be made according to whether the result is predictable or not.

References

Atabekov A, Yastrebov O (2018) Legal status of artificial intelligence across countries: legislation on the move. Eur Res Stud J 11(4):773–782

Bacaksız P, Sümer SY (2021) Robotlar, Yapay Zekâ ve Ceza Hukuku. Adalet Yayınevi, Ankara

Bak BB (2018) Medeni Hukuk Açısından Yapay Zekânın Hukuki Statüsü ve Yapay Zekâ Kullanımından Doğan Hukuki Sorumluluk. Türkiye Adalet Akademisi Dergisi 9(35):211–232

Bellman RE (2015) An introduction to artificial intelligence: can computers think? (quoted Hallevy G (2015) Liability for crimes involving artificial intelligence systems. Springer, p 6

Brozé B (2017) Troublesome 'Person'. In: Kurki VAJ, Pietrzykowski T (eds) Legal personhood: animals, artificial intelligence and the unborn. Law and philosophy library, vol 119. Springer International Publishing AG, pp 3–14
Centel N (2016) Ceza Hukukunda Tüzel Kişilerin Sorumluluğu – Şirketler Hakkında Yaptırım Uygulanması. Ankara Üniversitesi Hukuk Fakültesi Dergisi 65(4):3313–3326
Değirmenci O (2020) The crime of preparatory acts intended to disable protective programs in Turkish law (5846 Numbered Act Art. 72). In: 5. Türk – Kore Ceza Hukuku Günleri, Karşılaştırmalı Hukukta Ekonomik Suçlar Uluslararası Sempozyumu Tebliğler, C. II. Seçkin Yayıncılık, pp 1301–1322
Değirmenci O (2021) Yapay Zekâ ve Ceza Sorumluluğu. J Ardahan Bar 2(2):74–88
Doğan M (2021) Yapay Zekâ ve Özgür İrade: Yapay Özgür İradenin İmkânı. TRT Akademi 6(13):788–811
Downing R (2005) Shoring up the weakest link: what lawmakers around the world need to consider in developing comprehensive laws to combat cybercrime. Colum J Transnatl Law 705:705–762
Dyson GB (1997) Darwin among the machines: the evolution of global intelligence. Helix Books, p 7
Esin ME (2019) Yapay Zekânın Sosyal ve Teknik Temelleri. In: Telli G (ed) Yapay Zeka ve Gelecek. Doğu Kitabevi, pp 110–138
Hallevy G (2013) When robots kill, artificial intelligence under criminal law. Norheastern University Press
Hallevy G (2015) Liability for crimes involving artificial intelligence systems. Springer International Publishing
Hallevy G (2021) AI vs. IP, criminal liability for intellectual property offences of artificial intelligence entities. In: Baker DJ, Robinson PH (eds) Artificial intelligence and the law cybercrime and criminal liability. Routledge, pp 222–246
Hubbard FP (2022) Do androids dream?: Personhood and intelligent artifacts. https://papers.ssrn.com/sol3/papers.cfm?abstract_id=1725983. Accessed 02 March 2022
Jaynes TL (2020) Legal personhood for artificial intelligence: citizenship as the exception to the rule. AI & Soc 35:343–354
Kılıçarslan SK (2019) Yapay Zekânın Hukuki Statüsü ve Hukuki Kişiliği Üzerine Tartışmalar. J Yıldırım Beyazid Üniv Law Fac 4(2):363–389
Korkmaz İ (2018) Cihaz, Program, Şifre ve Güvenlik Kodlarının Bilişim Suçlarının İşlenmesi Amacıyla İmal ve Ticareti Suçu. Terazi Law J 13(142):45–55
Luzan T (2020) Legal personhood of artificial intelligence. Master's Thesis, University of Helsinki Faculty of Law
Miailhe N, Hodes C (2017) The third age of artificial intelligence. Retrieved February 3, 2022, from Special Issue 17, 2017: Artificial intelligence and robotics in the city. https://journals.openedition.org/factsreports/4383#tocto2n4
Michalczak R (2017) Animals' race against the machines. In: Kurki VAJ, Pietrzykowski T (eds) Legal personhood: animals, artificial intelligence and the unborn. Law and philosophy library, vol 119. Springer International Publishing AG, pp 91–101
Negri SMCA (2021) Robot as legal person: electronic personhood in robotics and artificial intelligence. Front Robot AI 8:1–10
Nilsson NJ (2018) Yapay Zekâ Geçmişi ve Geleceği, Mehmet Doğan (translator), Boğaziçi Üniversitesi Yayınevi, 19
Nowik P (2021) Electronic personhood for artificial intelligence in the workplace. Comput Law Secur Rev 42:1–14
Özen M (2003) Türk Ceza Kanunu Tasarısının Tüzel Kişilerin Ceza Sorumluluğuna İlişkin Hükümlerine Bir Bakış. Ankara Üniversitesi Hukuk Fakültesi Dergisi 52(1):63–88
Say C (2018) 50 Soruda Yapay Zekâ, 6. Baskı, İstanbul
Taşdemir Ö, Özbay ÜV, Kireçtepe BO (2020) Robotların Hukuki ve Cezai Sorumluluğu Üzerine Bir Deneme. J Ankara Üniv Law Fac 69(2):793–833
Tegmark M (2019) Yaşam 3.0 Yapay Zekâ Çağında İnsan Olmak. Pegasus Yayınları

Uzun FB (2016) Gerçek Kişilerin Hak Ehliyeti ve Hak Ehliyetine Uygulanacak Hukukun Tespiti". J Hacet Univ Law Fac 6(2):11–48
Winston PH (1992) Artificial intelligence, 3rd ed

Olgun Değirmenci He graduated from Istanbul University Faculty of Law in 1995. He completed his master's degree in criminal law with his thesis on Cyber Crimes in 2002, and his doctorate in criminal law at Marmara University in 2006 with his thesis on Laundering of Asset Values Arising from Crime. He became an associate professor of criminal and criminal procedure law in 2015 with his work on Numerical Evidence in Criminal Procedure. In 2020, he became a professor in the department of criminal and criminal procedure law. Since 2016, he has been working as a faculty member in criminal and criminal procedure law at TOBB University of Economics and Technology, Faculty of Law.

Chapter 7
Prevention of Discrimination in the Practices of Predictive Policing

Murat Volkan Dülger

Abstract Artificial Intelligence (AI)—driven regulations become increasingly prevalent in the field of law. In this context, the predictive policing practices used in detecting and preventing the potential criminality emerge. In the practices of predictive policing, basically, the data on the crimes committed in the past (such as the setting of the crimes—place and time, perpetrator and victim) are analyzed by algorithms, and a risk assessment regarding the commission of a new crime is conducted. Within the framework of the conducted risk assessment, the police take the necessary precautions to prevent the criminality. The use of AI in the practices of predictive policing creates an impression that risk assessment and its results are independent of human bias. On the contrary, AI algorithms involve the biases of the people who design the algorithms. In addition, the data analyzed by algorithms are not free from the biases of societies and class inequalities, either. Therefore, the practices of predictive policing are also likely to reflect discriminatory thoughts and practices of both individuals and societies. To prevent this situation, the practices of predictive policing should be critically examined, and possible solutions should be discussed. In this study, the side of predictive policing practices which is prone to discrimination will be examined, and the steps that can be taken to protect minorities and vulnerable groups in society will be discussed.

Keywords Predictive policing · Prohibition of discrimination · Artificial intelligence (AI) · Prevention of criminality

JEL Codes K14 · K24 · K38 · K42

M. V. Dülger (✉)
Faculty of Law, İstanbul Aydın University, Istanbul, Turkey
e-mail: volkan.dulger@dulger.av.tr

M. Kılıç and S. Bozkuş Kahyaoğlu (eds.), *Algorithmic Discrimination and Ethical Perspective of Artificial Intelligence*, Accounting, Finance, Sustainability, Governance & Fraud: Theory and Application, https://doi.org/10.1007/978-981-99-6327-0_7

7.1 Introduction

Artificial intelligence (AI) technology is widely used in different aspects of human activities in parallel with the developments in technology. The enlargement in the use of AI is changing our lifestyle including phones, medical care, finance, etc. (Rigano 2019). The influence of AI in our lives has also become apparent in the context of crime and criminal justice.

On the one hand, AI has varied the tools and techniques used to commit crimes. The occurrence of cybercrimes has improved by new methods such as AI-based cyber-attacks, AI-authored fake news, disrupting AI-controlled systems, burglar bots, AI-assisted stalking, and blackmailing people with "deep fake" technology (Caldwell et al. 2020). AI also has provided new methods for those who are not directly in cyberspace. The use of autonomous drones for attacks, military robots, and driverless vehicles as weapons can be given as examples of how AI has changed the tools for committing non-cyber crime (Caldwell et al. 2020).

On the other hand, AI has also changed this field by creating new opportunities to identify criminals, crimes committed (Dupont et al. 2018), and functioning of the criminal justice system[1] (Dammer and Albanese 2013). The use of AI in crime prediction and prevention is a contemporary development. It is almost actualized situation of the science fiction movie Minority Report. AI technology provides efficiency compared to human work and makes a difference with speed. Therefore, it is used to implement algorithms on large datasets to support or modify the law enforcement agencies (Yu et al. 2020). These adhibitions lead to practices of predictive policing by using AI and in this context, criminal and crime prediction.

7.2 Predictive Policing

Practices of predictive policing can be defined as "implementation of analytical techniques—particularly quantitative techniques—to identify likely targets for police intervention and prevent crime or solve past crimes or catch perpetrators red-handed by making statistical predictions" (Kreutzer and Sirrenberg 2020; Perry et al. 2013). In other words, practices of predictive policing, analyzing available data to predict where a crime might be committed in each time period, or who might be the victim or perpetrator of that crime (Richardson et al. 2019). Prediction of crimes by means of practices of predictive policing enable the organs of law enforcement agencies, especially the police, to act more efficiently and systematically. Consequently, it is aimed that the society has a more peaceful and safe community (Alkan and Karamanoğlu 2020).

[1] Criminal justice system refers to "to all institutions whose purpose is to control crime". It composed of police, courts and correctional institutions that works for interesting with criminals who are sentenced, judging suspects and enforcing the law.

Practices of predictive policing include estimating possible crimes as well as the location, time, perpetrator, and victim of these crimes. The crime scene can be determined as large as the entire area or as small as the corner of a particular street. Similarly, the time of crime can be determined at different times of the day, such as a day, a week, or a month. Both temporal units (the time of a crime) and spatial units (the location of a crime) are defined in accordance with social conditions (Berk 2021). For example, one estimate could be related to "a large increase in assaults and burglary in a particular block on a Saturday night immediately after bars close" or "an increase in break-in into jewelry stores on long holiday weekends" (Berk 2021). Criminal justice organs, especially the police, can take more precise measures to prevent crime through these estimates, which highlight the "black spots" where most crimes are committed.

In this respect, it is possible to divide practices of predictive policing into two categories: Offender-based modeling and geospatial modeling. Offender-based modeling relies on demographics and offender biographies to create risk profiles for individuals. In geospatial modeling, crime data is used to create risk profiles in urban areas (Shapiro 2017).

Statistics have been used in the field of criminology for decades to predict the occurrence of a crime. In addition to that, information technologies are increasingly being used to collect, store, and evaluate datasets to predict crime (Perry et al. 2013). The latest trend in this field is that instead of relying on the predictions of law enforcement employees, using AI in practices of predictive policing that harness the power of big data.

7.3 Use of Artificial Intelligence in Predictive Policing

Since there is no consensus on the definition of AI, it can be defined as "the ability to correctly interpret the external data of a system, learn from these data and use these learnings to achieve specific goals and tasks through flexible adaptation" (Kaplan and Haenlein 2019). In other words, AI can autonomously distinguish its environment, react, and without direct human intervention, it can perform tasks classically completed through human intelligence and decision-making ability (Rigano 2019). AI is far ahead of human work in terms of efficiency and speed (Yu et al. 2020); and also generally accepted that the capacity of AI to overcome (Rigano 2019) human errors is encouraging.

Although AI is being used more and more frequently in various fields, its use in the criminal justice system has a particular importance. Under the roof of criminal justice system, states are obliged to fulfill their responsibilities to ensure and secure public order, as well as to prevent rights violations against society through the detection of crimes, investigation, prosecution, and punishment. Accordingly, states have significant mandatory powers under their own laws, such as surveillance, detention, search, seizure, take into custody, and use of physical and even lethal force under certain circumstances. Suspects and perpetrators are given certain rights that limit these powers and prevent arbitrariness.

At this point, AI, to prevent criminal activity and ensure public safety, including "identify individuals and their actions in videos related to DNA analysis, firearm detection and crime prediction" (Rigano 2019) incorporated into the criminal justice system to address existing needs more quickly and easily. In video and image analysis, AI has great potential in matching faces, detecting objects such as weapons, and detecting the occurrence of an event. In forensics, it is much easier to study biological materials and create DNA profiles with AI-based techniques. In firearm analysis, AI helps determine the type (class and caliber) of weapons (Rigano 2019). In addition to the aforementioned uses of AI, it is also used in crime prediction and practices of predictive policing as mentioned in this proceeding. However, this use has disadvantages as well as advantages; especially, the main source of these disadvantages is the presence of data that can create bias in the big data pool that teaches and develops AI. Therefore, the inclusion of AI in the criminal justice system should be treated with special care when compared to other aspects of life.

As practices of predictive policing use data to predict crime and prevent future crime, AI is also being incorporated into practices of predictive policing to deal with large and complex datasets. Though the speed and efficiency of AI, the need for human studies in the analysis of crime prediction can be reduced.

AI-oriented algorithms used in the criminal justice systems of different countries are used with various variations. For example, AI applications have been used to analyze data on crimes committed in the past. These applications make risk assessments for future crimes after analyzing the data with algorithms. These assessments are used in the policymaking of public authorities in the fight against crime.

In Western countries, AI is used extensively in the field of security. One of the obvious examples is the use of AI in the field of predictive policing and aims to predict when and where crime is most likely to occur (Kreutzer and Sirrenberg 2020). During practices of predictive policing, data from past crimes are evaluated and information is given about the location, time, and nature of the crime. For this reason, this practice is called the "near-repeat theory" (Kreutzer and Sirrenberg 2020). For example, data such as address, time, occurrence of the case, apprehension from past theft cases, and population density or sociodemographic data such as building structure are collected and brought together in predictive policing software. This software shows high-risk areas and possibilities for new crimes. Law enforcement agencies are asked to conduct different inspections, such as patrolling the areas indicated by the software, secretly checking areas, or informing local residents.

An example of one of the current law enforcement agency activities of AI is the assignment of an AI software as a police officer in Dubai in 2017. In 2030, the first fully automatic police station is planned to be implemented. The police working in this station will aim to catch the perpetrators red-handed with the help of automatic payment processing, virtual assistants, and algorithms for the purpose of predicting and preventing crime. In addition to these, it will be possible to pay fines on these robots. With this application, it is aimed to create a quarter of the police force in Dubai (Hansford 2017).

Similarly, in the UK, Durham police are planning to use AI to score crimes in order to assess their probability of committing crimes. It is seen that the name of the evaluation tool will be HART (Harm Assessment Risk Tool) (Walker 2016; Gutwirth et al. 2015). As a matter of fact, Durham Police have been using this software since 2017 to determine who should and should not be arrested (BBC).

In another example, in 2016, the San Francisco Supreme Court began using an AI called PSA to determine whether criminals should be granted conditional release (Simonite 2017).

After the September 11 terrorist attacks, AI again promoted the collaborative use of criminal intelligence for crime prevention through a program called the US IPL (Intelligence-led policing). This program commits measurable results to reduce crime (Beck 2012).

Another example of such AI-driven technologies was created by California-based predictive law enforcement agency known as PredPol (The Predictive Policing Company). PredPol predicts where crimes can occur and based on its predictions, calculates the optimal allocation of police resources. Then it uses historical data from two to five years of the police department using the system to train a machine learning algorithm that is updated daily. Only three data points are used for this: type of crime, location, and date/time. No demographic, ethnic, or socio-economic information is used in PredPol. This eliminates the possibility of privacy or civil rights violations seen in other intelligence-driven or predictive policing models. However, it is stated that this claim of the company is controversial (Chace 2018).

PredPol was developed in the 1990s by criminologists who found a distribution model in home burglaries. This algorithm has been developed with the idea that previously entered houses are more likely to be re-entered and nearby houses are at a higher risk of being victimized by the same burglary. However, not all crimes can be followed with the model developed by PredPol. While burglaries are tracked with this software, crimes such as robbery, assault, or rape cannot be tracked. Detection of recurring, undetected crimes leads the program to identify people who seem negatively are poor and socially.

HunchLab is a similar application to PredPol but takes a different approach. Rather than relying on a single criminological concept, HunchLab bases its modeling systems on as many data sources as possible. Also in this software, in every city where data is available, machine learning techniques are used to create "case-specific" models for each type of crime. This method, called "gradient boosting", which uses thousands of decision trees, allows the algorithm to capture which variables are most likely. Thus, the probability of detecting crime types increases more. In other words, instead of catching all types of crimes with a single criminological concept, the algorithm in HunchLab is able to distinguish which criminological concepts best represent which crime patterns. This can provide a more detailed picture of the relevant variables used to predict different crimes but may also limit our perception of the structural preconditions of crime (such as poverty or unequal access to social resources).

Another of the most well-known applications of predictive policing is the software called COMPAS (Correctional Offender Management Profiling for Alternative Sanctions), which is offered by a company called "Northpointe". This application is a risk assessment tool used to support the decisions of criminal justice bodies and provides services for a fee (Cortes and Silva 2021). This software assesses not only risk but also about two dozen "criminogenic needs" related to theories of guilt, including "criminal personality", "social isolation", "substance abuse", and "residence/stability". Perpetrators are rated as low, moderate, or high risk in each category. The police take the necessary measures to prevent crime within the framework of the risk assessment made using this software. As can be seen, there are many AI-oriented applications, such as the examples we have mentioned for predictive policing applications.

This type of scoring, known as risk assessment for committing a crime, is increasingly used in courts across the United States. For example, as in Fort Lauderdale, they are used at every stage of the criminal justice system, from determining bail amounts to more fundamental decisions regarding the freedom of perpetrator. Thus, the decisions about who can be released within the criminal justice system are affected by the scores that this software creates about individuals. In some states of the United States, such as Arizona, Colorado, Delaware, Kentucky, Louisiana, Oklahoma, Virginia, Washington, and Wisconsin, the results of such assessments are given to judges during criminal sentencing. (Angwin et al. 2016).

In the United States, the results of such programs are used to rank a perpetrator's future risk of committing a crime, and generally to assess the perpetrator's rehabilitation needs. The Justice Department's National Institute of Corrections encourages the use of such combined assessments at all stages of the criminal justice process. Sentencing Reform Bill, passed by the U.S. Congress and enacted on April 13, 2022, mandates the use of such assessments in federal prisons (see also: https://www.congress.gov/bill/114th-congress/senate-bill/2123/text; http://uscode.house.gov/view.xhtml?req=(title:21%20section:801%20edition:prelim).

In 2014, then U.S. Attorney General Eric Holder warned that the risk scores might be injecting bias into the courts and he called for the U.S. Sentencing Commission it to examine the use of these software: "*Although these measures were crafted with the best of intentions, I am concerned that they inadvertently undermine our efforts to ensure individualized and equal justice; they may exacerbate unwarranted and unjust disparities that are already far too common in our criminal justice system and in our society.*" he said (Angwin et al. 2016).

Consequently, the success of practices of predictive policing will depend on how reliable they are, how different sources of information are integrated, and how all data is analyzed (Pearsall 2010).

Through practices of predictive policing, decisions about who to imprison will be made more effectively by predicting who is likely to commit crimes in the future. Besides, the use of overcrowded prisons can be optimized in this regard (Boobies 2018).

However, it can be criticized that AI interferes with the freedom of decision. A study from Stanford University reported that AI can do more than face recognition. In addition to facial recognition, it has been stated that there are AIs capable of not recognizing whether a person is heterosexual or homosexual (Levin 2017). A professor of Stanford, who carried out the research, recently announced that this AI will be able to detect people's susceptibility to delinquency.

In addition, it is not easy for the perpetrators to oppose the content of these software. As a result of the risk assessment and scoring made by this software, it is decided that the accused is not suitable for alternative execution methods to imprisonment—especially during the hearing of a criminal case—may turn into an imprisonment. At this stage, perpetrators rarely have an opportunity to object their assessment. Results are often shared with the perpetrator's attorney, but calculations that convert baseline data into points rarely occur. In other words, there is a "black box" in terms of evaluation criteria. Therefore, says Christopher Slobogin, director of Vanderbilt Law School's Criminal Justice Program, USA, "Risk assessments should be impermissible unless both parties get to see all the data that go into them; it should be an open, full-court adversarial proceeding" (Angwin et al. 2016).

After all, the use of AI in practices of predictive policing has unfortunately not been able to avoid prejudices, although it is an algorithm that will facilitate the pre-investigation and prosecution phase. AI algorithms contain the prejudices of the people who designed them. At this point, there is a violation of the prohibition of discrimination.

7.4 Prohibition of Discrimination

The prohibition of discrimination is the protection of individuals by the legal order in order to prevent their rights from being restricted or lost due to their physical and biological characteristics without the possibility of choice at birth (Doğan Yenisey 2005).

Article 14 of the European Convention on Human Rights ("**ECHR**") regulates the prohibition of discrimination by saying; "*The enjoyment of the rights and freedoms set forth in this Convention shall be secured without discrimination on any ground such as sex, race, colour, language, religion, political or other opinion, national or social origin, association with a national minority, property, birth or other status*". This clause in the Convention is complemented by additional protocols and other normative provisions (Yıldız 2012).

In addition, the general prohibition of discrimination is also regulated in Article 1 of Protocol No. 12 of the ECHR.

According to this Article;

1. In the case of a conflict between international agreements, duly put into effect, concerning fundamental rights and freedoms and the laws due to differences in provisions on the same matter, the provisions of international agreements shall prevail.
2. No one shall be discriminated against by any public authority on any ground such as those mentioned in paragraph 1.

Article 2 of the International Covenant on Civil and Political Rights, Article 2 of the International Covenant on Economic, Social and Cultural Rights, and the African Charter on Human and Peoples' Rights also include provisions on the prohibition of discrimination (Temperman 2015). International Covenant on Economic, Social and Cultural Rights generally prohibits any act that incites states parties to discrimination or violence (Temperman 2015).

Turkey is a party to the conventions regulating the prohibition of discrimination. Article 90 of the Constitution also states that in cases where the dispute comes before the judiciary, the relevant provisions of the international human rights conventions prohibiting discrimination will be applied (Karan 2007):

> ...In the case of a conflict between international agreements, duly put into effect, concerning fundamental rights and freedoms and the laws due to differences in provisions on the same matter, the provisions of international agreements shall prevail.

In this respect, the prohibition of discrimination from a normative point of view is accepted as a basic principle in Turkish law (Uyar 2006). As a matter of fact, the right not to be discriminated against as a regulated right in the Constitution has been made functional by turning it into a prohibition norm, that is, a type of crime, with the Turkish Penal Code. Thus, the right not to be discriminated against is regulated by the TPC as a legal value protected by crime.

The regulation regarding the prohibition of discrimination in Article 122 of the Turkish Penal Code ("**TPC**") is as follows:

> Any person who
>
> a) Prevents the sale, transfer or rental of a movable or immovable property offered to the public,
> b) Prevents a person from enjoying services offered to the public,
> c) Prevents a person from being recruited for a job,
> d) Prevents a person from undertaking an ordinary economic activity on the ground of hatred based on differences of language, race, nationality, colour, gender, disability, political view, philosophical belief, religion or sect shall be sentenced to a penalty of imprisonment for a term of one year to three years.

The prohibition of discrimination includes the right to equality and equal protection. The principle of equality is expressed as a positive obligation in this context (Ramcharan 1981). In order to fulfill this positive obligation, this prohibition has been regulated both as a fundamental right and freedom in the Constitution and as a crime norm in the TPC, and it has been clearly stated that those who violate this prohibition will be punished. However, in Turkey, we do not have any concrete data beyond an abstract opinion on how this abstract norm protection works based on concrete events.

7.5 The Use of Artificial Intelligence in Predictive Policing and the Prohibition of Discrimination

The use of AI in the public and private sectors has caused significant concerns from various aspects, including technical issues, ethics (Berk 2021) and discrimination (Dupont et al. 2018). Firstly, AI algorithms and big data work in a way that is invisible to the public, sometimes even to their creators (the black box phenomenon we mentioned above) (Leurs and Shepherd 2017). Secondly, AI is pretty good at processing data and drawing conclusions from it; however, if the available data has a biased nature, AI reproduces this bias (Dupont et al. 2018). They maintain existing discriminatory practices because the algorithmic operations of AI are not public and are exempt from investigation (Pasquale 2015; Leurs and Shepherd 2017). In this proceeding, these discriminatory practices will be briefly exemplified under the categories of race and gender.

There has been a lot of research showing how racial ideology affects AI concepts. In these studies, it was emphasized that while images of "prestigious business" such as medical doctors or scientists were depicted as white, Latinos were depicted as professionals in the porn industry (Noble 2018). Similarly, humanoid robots are mostly designed as white with blue eyes and blonde hair. Virtual assistants like Siri can speak in some special accents but not African-American English (Cave and Dihal 2020). All these examples, in the end, Cave and Dihal (2020) rightly state that "it has been empirically shown that machines can be racialized, in this context, that machines can be given qualities that will enable them to identify with racial categories".

A lot of research has been done on how gender biases shape AI technologies. The most prominent of these is the report of UNESCO, which includes its findings on this issue. In this report, it is clearly stated that the biases of gender discrimination, especially in terms of women and LGBT people, are also transferred to AI software through big data and this leads to discrimination (UNESCO 2020).

Not surprisingly, AI technologies used in the criminal justice system reproduce the same biases. Many studies have shown that COMPAS grapples with racial biases.

When predictive AI algorithms are trained on biased datasets, existing biases are reproduced (Hayward and Maas 2020). For example, ProPublica checked the records of 7,000 people arrested in the United States in 2013 and 2014 and conducted an extensive investigation into practices of predictive policing. As a result of this research, it was determined that only 61% of the total prisoners are likely to commit crimes again; it was stated that the algorithm used in the cases examined was specific and was made considering previous criminal records (Boobies 2018; Angwin et al. 2016).

Many problems arise due to the use of AI in the field of predictive policing. One of these problems is that the stored information and the generated risk score can be manipulated. Even at the learning stage of AI, law enforcement agency can decide which data can be a resource of the algorithm. Thus, people's ethnic origins, gender or religious views can become discriminatory and pose a danger in this respect. In this way, discriminatory qualifications can be added to the software, forming the basis for future law enforcement agency measures. In this way, "a bias" can be introduced into software not only consciously but also unconsciously. For example, In the United States, it has been seen that the law enforcement agency activities which with the help of AI take more black and poor people into their suspicious circles in their foresight-based activities. Thus, people in this group are questioned and punished more than usual. Hereby, social inequality and racial prejudice are protected and sustained within the legitimacy of science (Karakurt 2019).

Another problem is that in practices of predictive policing made with the help of AI, algorithms are exposed to their own acceleration during their use and constantly change themselves as soon as new information is added. This causes a paradox that makes it impossible to predict algorithmic decision-making processes. In this way, the software becomes a black box and its internal processes become opaque. With a lack of transparency, it is not possible to understand what properties the algorithm uses to calculate a result. Therefore, it is difficult to confirm the results. Discrimination can become noticeable only after many cases are processed by an algorithm (Karakurt 2019). The COMPAS software, which we cited as an example and used in the United States, also measured individuals' risk of being accused and found that blacks were more likely to be accused. This result was achieved even if the skin color of the persons was not recorded as data at the beginning of the software. Here, too, the "anti-discrimination phenomenon" arising from the impression of objective and unbiased software has disappeared (Karakurt 2019). The disappearance of this prediction and the direct trust in the software lead to the emergence of discriminatory, unjust, and unlawful results and practices.

Whether the data can be used in terms of the Constitution is another problem. Technically, the total surveillance of all data, the creation of closed personality profiles, or the use of tools such as face recognition without any reason may lead to a violation of the rights in the Constitution (Karakurt 2019). This may create a violation of both the regulation on the prohibition of discrimination in the Constitution and the right to protection of personal data regulated in paragraph 3 of Article 20.

As we mentioned earlier PredPol and HunchLab systems have advantages as well as disadvantages. A study by members of the Human Rights Data Analysis Group found that, when these are applied to drug-related crimes, it has been found that police officers are sent to areas with high ethnic minority populations for drug inspections, despite evidence of an even distribution of data on drug use across regions.

PredPol finders argue that software like HunchLab triggers discrimination more. The reason for this is that systems such as HunchLab store more data, causing them to associate ethnic origins much more. There is currently no concrete evidence as to which system leads to more discrimination. Law enforcement agencies receive subsidies from government funds to carry out practices of predictive policing. However, law enforcement agencies are not obliged to control how these softwires affect social equality. There are many preventive (predictive) law enforcement practices currently used in many countries of the world, but their legitimacy is debatable. The use of such software is increasing, especially in countries with more authoritarian governments. In addition, it should not be ignored that these programs have commercial purposes and may become the sole activity of law enforcement agencies (Shapiro 2017).

7.6 What the Future of Turkey Tells Us?

Crimes that are committed every day in our country place extra burden on institutions such as law enforcement agency and prosecution. This situation has a negative effect on the justice system from beginning to end. While it causes the prolongation of the trial (investigation and prosecution) process in the first place, this prolonging process affects the public conscience and undermines trust in justice (Alkan and Karamanoğlu 2020).

As far as we know, AI-oriented technology is not yet used in crime prediction and practices of predictive policing in Turkey. There is also no legal regulation regarding this. However, this does not mean that Turkey will not use such technologies in its criminal justice system in the future. Discussions about the possible use of such technologies will raise the attention and awareness of both policymakers and society.

Therefore, such technologies, which are highly likely to come to our country in the future, advantages and disadvantages should be talked and discussed already. Especially policymakers, sociologists, philosophers, criminologists, and lawyers should take part in these discussions, and people in these groups should be provided to freely share their views. Otherwise, (as has been practiced up to the present) the introduction of such practices without these discussions will lead to a further increase in discrimination based on various criteria, which is already excessive in our country. This will lead to inequality, the deterioration of legal security and order, and as a result, the deterioration of social peace. It should not be forgotten that societies in which trust in the law has decreased or disappeared cannot remain standing for a long time. For this reason, the rules of law should be applied equally to everyone without any discrimination. This practice in force for the use of AI software for the prevention and/or repetition of crime.

References

Alkan N, Karamanoğlu YE (2020) Öngörüye Dayalı Kolluk Temelinde Önleyici Kolluk: Rusya Federasyonu'ndan Örnekler. Güvenlik Bilimleri Dergisi

Angwin J, Larson J, Mattu S, Kirchner L (2016) Machine bias: there's software used across the country to predict future criminals and it's biased against blacks. https://www.propublica.org/article/machine-bias-risk-assessments-in-criminal-sentencing.

Beck C (2012) Predictive policing: what can we learn from wal-mart and amazon about fighting crime in a recession? https://www.policechiefmagazine.org/predictive-policing-what-can-we-learn-from-wal-mart-and-amazon-about-fighting-crime-in-a-recession/

Berk R (2021) Artificial intelligence, predictive policing, and risk assessment for law enforcement. Ann Rev Criminol 209–237

Boobies T (2018) Analytics for insurance. The real business of big data. Wiley Finance Series

Caldwell M, Andrews JTA, Tana T, Griffin LD (2020) AI-enabled future crime, crime science. Council of Europe Committee on Legal Affairs and Human Rights Report. https://doi.org/10.1186/s40163-020-00123-8

Cave S, Dihal K (2020) The whiteness of AI. In: Philosophy and technology, vol 33. Springer Verlag

Chace C (2018) Artificial intelligence and the two singularities. CRC Press, New York

Cortes ALL, Silva CF (2021) Artificial intelligence models for crime prediction in urban spaces. Mach Learn Appl Int J (MLAIJ)
Dammer HR, Albanese JS (2013) Comparative criminal justice systems. Cengage Learning
Doğan Yenisey K (2005) Eşit Davranma İlkesinin Uygulanmasında Metodoloji ve Orantılılık. Legal İHSGH Dergisi
Dupont B, Stevens Y, Westermann H, Joyce M (2018) Artificial intelligence in the context of crime and criminal justice. Report for the Korean Institute of Criminology
Gutwirth S, Leenes R, De Hert P (2015) Data protection on the move. Springer Verlag
Hansford M (2017) Mühendisler neden dijital demiryolundan heyecan duymalı? Yeni İnşaat Mühendisi
Hayward KJ, Maas MM (2020) Artificial intelligence and crime: a primer for criminologists
Karan U (2007) Türk Hukukunda Ayrımcılık Yasağı ve Türk Ceza Kanunu'nun 122. maddesinin Uygulanabilirliği. TBB Dergisi, Sayı 73
Kaplan AM, Haenlein M (2019) Siri, Siri, in my hand: who's the fairest in the land? On the interpretations, illustrations, and implications of artificial intelligence. Business Horizons
Karakurt B (2019) Predictive policing und die Gefahr algorithmischer Diskriminierung. Humboldt Law Clinic
Kreutzer RT, Sirrenberg M (2020) Understanding artificial intelligence fundamentals, use cases and methods for a corporate AI journey. Springer
Leurs K, Shepherd T (2017) Datafication and discrimination. https://www.degruyter.com/document/doi/10.1515/9789048531011-018/html, 14.04.2022
Levin S (2017) New AI can guess whether you're gay or straight from a photograph. https://www.theguardian.com/technology/2017/sep/07/new-artificial-intelligence-can-tell-whether-youre-gay-or-straight-from-a-photograph
Noble SU (2018) Algorithms of oppression: how search engines reinforce racism. New York University Press
Pearsall B (2010) Predictive policing: the future of law enforcement? Natl Inst Justice J
Perry WL, McInnis B, Price CC, Smith SC, Hollywood JS (2013) Predictive policing: the role of crime forecasting in law enforcement operations. RAND Corporation
Pasquale F (2015) The black box society: the secret algorithms that control money and information by Frank Pasquale. Harvard University Press, Cambridge
Ramcharan B (1981) Equality and non-discrimination. In: The international bill of rights: the covenant on civil and political rights. Columbia University Press
Rigano C (2019) Using artificial intelligence to address criminal justice needs. Nat Inst Justice J
Richardson R, Schultz JM, Crawford K (2019) Dirty data, bad predictions: how civil rights violations impact police data, predictive policing systems, and justice. New York University School of Law
Shapiro A (2017) Reform predictive policing. Nature 541:458–460
Simonite T (2017) When government rules by software, citizens are left in the dark. https://www.wired.com/story/when-government-rules-by-software-citi-zens-are-left-in-the-dark/
Temperman J (2015) Religious hatred and international law: the prohibition of incitement to violence or discrimination. Cambridge University Press
Uyar L (2006) Birleşmiş Milletler'de İnsan Hakları Yorumları: İnsan Hakları Komitesi ve Ekonomik, Sosyal ve Kültürel Haklar Komitesi 1981–2006. Bilgi Üniversitesi Yayınları
UNESCO (2020) Artificial intelligence and gender equality: key findings of UNESCO's global dialogue
Walker PJ (2016) Croydon tram driver suspended after video of man 'asleep' at controls. The Guardian
Yu H, Liu L, Yang B, Lan M (2020) Crime prediction with historical crime and movement data of potential offenders using a spatio-temporal Cokriging method. ISPRS Int J Geo Inf
Yıldız C (2012) Avrupa İnsan Hakları Sözleşmesi'nin "Ayrımcılık Yasağını" Düzenleyen 14. Maddesinin, Avrupa İnsan Hakları Mahkemesi Kararlarıyla Birlikte İncelenmesi. İstanbul Kültür Üniversitesi Hukuk Fakültesi Dergisi

Murat Volkan Dülger He graduated from Istanbul University Faculty of Law in 2000. He received his master's degree from Istanbul University Institute of Social Sciences with his study on "Information Crimes in Turkish Criminal Law" in 2004, and his doctorate in law with his doctoral thesis on "Crimes and Sanctions Related to Laundering of Assets Resulting from Crime" in 2010. He works as a faculty member in Criminal and Criminal Procedure Law and IT Law Departments at Istanbul Aydın University Faculty of Law and as the head of both departments. He has focused his studies on Criminal Law, Criminal Procedure Law, IT Law, Personal Data Protection Law, Legal Liability of Artificial Intelligence Assets and Human Rights Law.

Chapter 8
Issues that May Arise from Usage of AI Technologies in Criminal Justice and Law Enforcement

Benay Çaylak

Abstract Due to the constant and swift technological advancements, artificial intelligence technologies have become an integral part of our daily lives and as a result, have started to impact various areas of our society. Legal systems proved to be no exception as many countries took steps to implement AI technologies to their legal systems in order to improve the law enforcement and criminal justice systems, making changes in various processes including but not limited to preventing crimes, locating perpetrators, accelerating judicial processes, and improving the accuracy of judicial decisions. While the usage of AI technologies provided improvements to criminal justice and law enforcement processes in various aspects, concerning instances demonstrated that AI technologies may reach to biased, discriminatory, or simply inaccurate conclusions that may cause harm to people. This realization becomes even more alarming considering that criminal justice and law enforcement consist of extremely critical and fragile processes where a wrong decision may cost someone their freedom, or in some cases, life. In addition to discrimination and bias, automated decision-making processes also have a number of other issues such as lack of transparency and accountability, jeopardization of the presumption of innocence principle, and concerns regarding personal data protection, cyber-attacks, and technical challenges. Implementing AI technologies to legal processes should be encouraged since criminal justice and law enforcement could benefit from recent advancements in technology and it is possible that more accurate, more just, and faster judicial processes can be created. However, it should be carefully considered that implementing AI systems which are in their infancy to legal processes that could lead to severe consequences may cause incredible and, in some cases, irrevocable damages. This study aims to address current and possible issues in usage of AI technologies in criminal justice and law enforcement, providing possible solutions when possible.

B. Çaylak (✉)
Istanbul University, Istanbul, Turkey
e-mail: benaycaylak@gmail.com

M. Kılıç and S. Bozkuş Kahyaoğlu (eds.), *Algorithmic Discrimination and Ethical Perspective of Artificial Intelligence*, Accounting, Finance, Sustainability, Governance & Fraud: Theory and Application, https://doi.org/10.1007/978-981-99-6327-0_8

Keywords Artificial intelligence · Criminal justice · Law enforcement · Discrimination · Bias · Transparency · Accountability · Algorithmic black box · Opaque decision making · Presumption of innocence · Protection of personal data · Cyberattacks · AI literacy

8.1 Introduction

Recent history has been a stage of continuous and incredible technological advancements. With popularization of the internet accelerating technology's integration to our lives, we were led to the times of artificial intelligence ("**AI**") and internet of things. Nowadays, it seems impossible to live a day without encountering a semi-automated or automated system. This fundamental change is advantageous for both individuals and the society as a whole, because a more digitally literate and technology-oriented society has the potential to constantly evolve and over time eliminate challenging and detrimental aspects of societal life.

Upon its rise in popularity and the increase in development and implementation of AI technologies, AI started to become a part of many sectors, starting from technology-oriented fields such as IT and e-commerce. However, it did not take long for other sectors to catch up with the advancements that were made possible by AI technologies.

Legal sector, despite being notoriously conventional, wasted no time in taking steps to include AI technologies in its processes as many countries acted upon implementing a number of AI technologies to their legal systems. Criminal justice and law enforcement were among the legal fields that were affected by this change. The main objectives of the developments in these fields being including but not limited to preventing crimes, locating perpetrators, accelerating the judicial processes, and improving the accuracy of judicial decisions.

However, implementing AI technologies in criminal justice processes and law enforcement proved to be a not-so-easy task, especially considering the very nature of criminal justice and law enforcement.

AI technologies have thus far proven to be revolutionary in the sense of bringing a more objective, unquestionable, and fast approach to the legal field, providing assistance regarding a number of existing issues in the field including but not limited to the "human error" factor, the ever-increasing workload that does not seem to show a downward trend anytime soon, the lack of objectivity in some cases and the horrifying outcome that is wrongful accusations, convictions and enforcement of undeserved sentences upon innocent people. Though AI processes are certainly not without pitfalls that unfortunately may cause serious damage.

This study aims to analyze the current and possible issues that may arise from using AI technologies in criminal justice and law enforcement, also providing possible approaches that may lessen or completely eliminate the negative outcomes.

Before the issues are elaborated on, please bear in mind that this study is actually pro-AI and does not aim to undermine the positive effects AI has on both individuals and societies. The prospects of AI being integrated even more into the legal field and changing the legal field and the world for the better is extremely exciting. However, these issues must be immediately and thoroughly discussed in order to prevent possible negative, even catastrophic consequences. This is the only way to ensure progress, take the best aspects of AI technologies, and leave out the worst.

The main issues that will be discussed in this study are, (i) discriminatory and biased decisions, (ii) issues surrounding transparency, accountability, and the right to a fair trial, (iii) issues surrounding the presumption of innocence, (iv) possible risks that may arise with regard to personal data protection and the data minimization principle, (v) the vulnerability of AI technologies against cyber-attacks, and (vi) technical aspects of AI implementation.

Moreover, this list of issues is not prioritization based, all of the issues that will be discussed below carry importance on their own merit, though their urgency and severity may vary in parallel with the circumstances of the concrete cases. Also, the list of issues that will be elaborated on in this study is by no means exhaustive.

8.2 Current and Possible Issues

8.2.1 Discrimination and Bias

In automated decision-making processes, AI technologies make decisions based on the data they are fed. Thus, the accuracy and quality of the fed data is a noteworthy factor that shapes the accuracy and quality of the output. The data, due to any reason, being tampered with, compromised, biased, or discriminatory would directly affect the decision. Biased and/or discriminatory data inevitably results in biased and/or discriminatory outputs (Abudureyimu and Oğurlu 2021: 774).

Since discriminatory and/or biased outputs pose a great threat to judicial processes, it is crucial to get to the bottom of the reasons that may cause this. At this point, it is important to note that the list of possible reasons that will be explained below is not exhaustive.

One of the possible reasons for discriminatory or biased outputs is prioritizing certain characteristics, features, or criteria in the decision-making process. This may lead to a bias toward the subjects that possess the prioritized characteristics/features/criteria. Oppositely, the subjects that lack the prioritized features may be discriminated against (Borgesius 2018: 20). Another possible reason would be the deficiencies that stem from the AI technology itself. The third possible reason is an external intervention to the AI technology such as data manipulation (İçer 2021: 41), data poisoning,[1] and cyber-attacks that tamper with the system or the data within the AI.

[1] In "**data poisoning**", a wrong data is deliberately included in order to create biased outputs. (European Parliament 2021, see par. P).

The final possible reason that will be listed in this study is "human error" transferring to AI technology. Decision-making processes of AI systems have for the longest time drawn comparisons to humans' decision-making. One of the biggest advantages of AI technologies is speculated to be lacking the "weaknesses" that humans have. To exemplify, since AI technologies do not get tired, stressed, or emotional, in some ways, they operate way better than humans. They are not human, so they do not make "human errors".

With that being said, especially considering that AI technologies still have a very long way to go in terms of developing, humans are still an integral part of the decision-making process. It is often overlooked that, for the time being, the information that is fed to AI technologies is provided by humans.

This is where human error presents itself in AI decision-making processes. Feeding an AI technology information that was a product of "human error" or "human bias" would only result in flawed and damaging outputs. Hence, the errors, biases, and discriminatory behavior that humans have eventually turn out to be the errors, biases, and discriminatory behavior that the AI has.

Biases of humans may be either deliberate or indeliberate. It is necessary to take a better look at these biases to deduct how they affect automated decision-making processes.

To start with indeliberate biases, it must be emphasized that since humans are imperfect, instinctive, and emotional by design; thoughts, tendencies, and decisions of humans heavily depend on biases they have such as (i) **cognitive**[2] **and perceptual biases**, which are biases that occur because of humans' genetics, upbringing, experiences, etc., (ii) **anchoring bias**,[3] which explains that humans tend to stick with the first information or judgement that they encounter and look for their "anchor" in other things, or (iii) **confirmation bias**,[4] which is simply trying to find evidence that supports preexisting thoughts in everything that is encountered (Schwartz et al 2022: 9).

As for deliberate biases, it is unfortunate to point out that deliberate discriminatory and/or biased behavior still occurs in so many levels and against so many people. Discrimination is sadly a current reality of the world we live in.

---

[2] Cambridge Dictionary defines "**cognitive bias**" as "the way a particular person understands events, facts, and other people, which is based on their own particular set of beliefs and experiences and may not be reasonable or accurate". (Cambridge) For an alternative definition, please see Schwartz et al. 2022: 49.

[3] American Psychological Association (APA) Dictionary defines "**anchoring bias**" as "the tendency, in forming perceptions or making quantitative judgments under conditions of uncertainty, to give excessive weight to the starting value (or anchor), based on the first received information or one's initial judgment, and not to modify this anchor sufficiently in light of later information". (American Psychological Association) For an alternative definition, please see Schwartz et al. 2022: 49.

[4] Brittanica defines "**confirmation bias**" as "the tendency to gather evidence that confirms preexisting expectations, typically by emphasizing or pursuing supporting evidence while dismissing or failing to seek contradictory evidence". (Brittanica) For an alternative definition, please see Schwartz et al. 2022: 50.

Apart from the situation described above, there is also the third possibility of the current data that is fed to AI technologies being biased by design and not due to humans' deliberate or indeliberate discriminatory actions such as the discriminatory and/or biased tendencies that are already inherent in underlying datasets, particularly in historical data (European Parliament 2021: par. Q. 8).

All things considered; it is still not possible to say that humans must not be involved in AI decision-making processes. Because, it has been detected before that AI technologies, without the supervision of humans, might act in a discriminatory or biased manner (Abudureyimu and Oğurlu 2021: 774). What must be focused on instead is making efforts in eliminating discriminatory and/or biased tendencies, decisions, and outputs in both humans and AI technologies separately and also when they act together.

Discriminatory and/or biased decision-making becomes even more concerning when we take into consideration the possible severe consequences that discriminatory and/or biased data and outputs may have in criminal justice and law enforcement, mostly due to the very nature of these fields.

Criminal justice and law enforcement are such important and delicate areas of law that discriminatory and/or biased decision-making can have utterly catastrophic consequences. These legal fields have the ability to literally change human lives. Their significance in human lives and society inevitably brings with itself a potential danger.

Decision-making processes affected by discrimination and/or bias, along with non-transparent thus not objectible decisions, may cause life-altering results such as wrongful decisions, convictions, and acquittals, resulting in failure to establish and maintain a just society.

A wrongful decision may cause someone to wrongfully spend 20 years of their life in prison despite being innocent. If someone is wrongfully imprisoned as a result of a discriminatory and/or biased decision-making process and if death sentence is still in force in the country the decision was made in, which is currently the case for many nations, an innocent person may be sentenced to death in vain and thus lose their life. What could ever be more severe than this?

The issues that were explained above makes discrimination and bias one of the, if not the most important issue that must be dealt with when it comes to AI in criminal justice and law enforcement.

It is obvious that eliminating or even mitigating this issue is not going to be easy. However, being mindful of the possibility of discriminatory and/or biased decisions, preventing and mitigating the risks for discrimination while paying special attention to groups that have an increased risk of being disproportionately affected by AI (Council of Europe Commissioner for Human Rights 2019: 11), executing the algorithmic decisions that create a new legal status or a significant change for people after the decision being reviewed by a human (Büyüksağiş 2021: 531) and in a legislative sense, making the necessary regulations in order to oversee current or possible pitfalls and ensuring compliance to these regulations are sure to go a long way.

8.2.2 *Issues Surrounding Transparency, Accountability, and the Right to a Fair Trial*

Upon conducting a decision-making process, it must be thoroughly explained as to why and how the decision was made and which criteria were considered through written, case-specific, and reasoned decisions. At the same time, the people whom the decision was made about, especially the people whom the decision was made against should be able to completely understand how the decision was made and, if it is in their favor to do so, object to it (Fair Trials 2021: 32).

What does actually happen between the start and the end of the AI decision-making process? This is a loaded question, and also one that seldom has a definite answer. Most of the time, not much can be understood regarding the ins and outs of the AI decision-making process, since we more often than not have no idea of the details. This is very a dangerous deficiency since it is crucial for the people affected by the decisions made by the AI technologies to be able to challenge these decisions in a meaningful and effective way (Demirtaş 2021: 6).

Transparency is the basis of fundamental rights such as right to a fair trial (Demirtaş 2021: 6). However, most of the time transparency is not granted to the people who the decisions are made about, due AI decision-making often being "opaque" (Borgesius 2018: 72). Transparency, which is a necessary part of the oversight (European Parliament 2021: par. Q. 17), cannot be possible when the decision-making is opaque.

Mysterious decision-making processes have been named as the "algorithmic black box".[5] Even the makers or operators of the AI most of the time do not know anything regarding what the algorithmic black box contains (Abudureyimu and Oğurlu, see p. 775; Fair Trials, see p. 33). This obviously poses a significant and worrying risk, making transparency very hard, if not entirely impossible.

If someone does not know exactly why the outcome is not in their favor, how can they respond and provide an effective objection that has even a slight chance of turning the decision around? How can someone defend themselves against something they don't know anything about? And if we don't know how and why a decision was made and which criteria were taken into account, how can we object to it and prove its inaccuracy? How do we challenge AI and hold it accountable? (Fair Trials, see p. 33).

Providing reasoning in legal processes is universally accepted due to the reasons explained above. Not providing a case-specific reasoning for a decision constitutes a violation of the right to a fair trial, which is a universal legal principle that was regulated in a variety of national and international legal provisions, particularly in Article 6 of the European Convention of Human Rights.

Algorithmic black box and opaque decision-making collectively result in a chain reaction of a non-transparent decision-making process, which produces a decision without a detailed and case-specific reasoning.

[5] For more information on "algorithmic black box", please see Bathaee 2018.

As a solution to this issue, we have to "demystify" the black box. We need to know how AI makes the decision, so we can ensure transparency. In this vein, striving toward having a better understanding of the AI decision-making process and conducting programming, operation, and maintenance processes in light of the need to understand what's in the "Black Box" would be ideal toward ensuring transparency. Furthermore, envisioning stricter and more objective standards for holding AI systems accountable is necessary.

8.2.3 Issues Surrounding Presumption of Innocence

Presumption of innocence, which can be defined pursuant to Article 48 of the European Charter of Human Rights as someone being charged being presumed innocent until they are proven to be guilty according to law, is a fundamental human right that is a core part of the right to a fair trial, envisioned in a number of legal documents such as the European Union Directive 2016/343 (Presumption of Innocence Directive) and the aforementioned Article 48 of the European Charter of Human Rights. (Fair Trials 2021: 30).

With regard to presumption of innocence, the main problem that poses a threat to criminal justice and law enforcement is preventive policing's dichotomy (crime prevention vs. jeopardizing presumption of innocence) and its relation to the principle of presumption of innocence.

Preventive policing, by concept, is contradictory to the principle of presumption of innocence since let alone concrete proof of being guilty, policing and assumption are made even before there is a committed criminal act and people are preemptively judged as guilty on the basis of their past behavior or other people they share similar characteristics and background (Fair Trials 2021: 30).

What can be done at this point is to establish and maintain a balance. Overseeing preventive policing activities and ensuring fair, non-discriminatory, non-biased and proportionate preventive policing processes would contribute greatly to the balance between these two concepts.

8.2.4 Protection of Personal Data Issues and Data Minimization

8.2.4.1 Protection of Personal Data

Legal processes, being thorough, detail-oriented, and evidence-based, by nature, include a large scale of data, both in the sense of quantity and variation. For example, a standard criminal case file consists of motions, witness depositions, hearing minutes, expert reports and any and all kinds of evidence that was submitted to the court or found upon ex-officio investigation. Digitalizing the court files and conducting criminal justice processes would be beneficial in terms of minimizing the use of paper and stationary, expediting the processes and eliminating the necessity of physical attending for process' actors such as witnesses (which turned out to be extremely crucial in light of the COVID 19 pandemic). However, these positive outcomes are not without challenges. Where there is a large source of data that is going through a digitization process, it is not possible not to be concerned regarding the safety and security of this data.

When it comes to ensuring the safety and security of the personal data that is involved in automated decision-making processes in a regulatory sense, Turkish law and European Union law have quite different approaches.

In Turkish Law, while there is no explicit regulation regarding direct implementation of decisions that were made with automated decision-making (Büyüksağiş 2021: 529, 532), AI technologies that base on personal data processing must be in compliance with Law number 6698 on Protection of Personal Data ("**KVKK**") and its sub legislation (Turkish Personal Data Protection Authority 2021:7).

Since Turkish Personal Data Protection Authority ("**the Authority**") is acutely aware that the link between AI and personal data protection must be assessed and regulated, in September of 2021, the Authority has published "***Recommendations Regarding Protection of Personal Data in the Field of Artificial Intelligence***".[6] These recommendations, which are in the same vein of guidelines that Council of Europe has, namely the "*Guidelines on Artificial Intelligence and Data Protection*",[7] is an adequate initial document for regulating this issue.

Throughout the recommendations in this document, which were grouped into sections in accordance with who they were aimed at ("general", "for developers, manufacturers and service providers" and "for decision-makers"), respecting a person's honor and fundamental rights, minimizing possible and current risks, and complying with national and international regulations are repeatedly emphasized (Yazıcıoğlu et al. 2022a, b).

The European Union, on the other hand, has an explicit provision that regulates this situation. In Article 22(1) of European Union General Data Protection Regulation ("**GDPR**"), it is regulated that; "*The data subject shall have the right not to be subject to a decision based solely on automated processing, including profiling, which produces legal effects concerning him or her or similarly significantly affects him or her*", unless the exceptions stated in paragraph (2) of the same Article[8] are the case. It was also envisaged in Article 22(3) that the data controller shall implement

[6] For the full Turkish text, please see https://www.kvkk.gov.tr/Icerik/7048/Yapay-Zeka-Alaninda-Kisisel-Verilerin-Korunmasina-Dair-Tavsiyeler (Last Accessed: 10.05.2022).

[7] For the full English text, please see https://rm.coe.int/guidelines-on-artificial-intelligence-and-data-protection/168091f9d8 (Last Accessed: 10.05.2022).

[8] Article 22(2) of GDPR is as follows.

"Paragraph 1 shall not apply if the decision:

1. is necessary for entering, or performance of, a contract between the data subject and a data controller;

suitable measures to safeguard the data subject's rights and freedoms and legitimate interests.

In addition to the legislative efforts showcased above, people involved in AI decision-making processes must also act in accordance with fundamental personal data protection principles. The right to protection of personal data and its basic principles such as lawfulness and fairness, data minimization, privacy-by-design, proportionality, accountability (European Parliament 2021: par. Q. 4), security and safety must be considered all throughout the processes (European Parliament 2021: par. Q. 1).

8.2.4.2 Data Minimization

Envisioned in Article 5(c) of GDPR as a personal data protection principle, data minimisation is defined as the personal data being "*adequate, relevant and limited to what is necessary in relation to the purposes for which they are processed*". To simplify, just because personal data can be collected and/or processed, does not mean it has to be. If it is not necessary, it must not be collected, kept, or processed.

What should be understood from "adequate, relevant and limited" depends on the purpose of data collection and usage. This means that, why a data is needed must be firstly discussed and decided on (Information Commissioner's Office (ICO), "**Data Minimization**"). However, this is not always possible in automated decision-making processes, if ever. Hence, a number of issues emerge with regard to data minimization (please bear in mind that this is not an exhaustive list):

a. In AI, the more data are collected, the more accurate the decision becomes. In this light, usually, as much data as possible, if not all available data, is collected (Dülger et al. 2020: p. 8). This approach is a direct and severe violation of the data minimization principle.
b. New data are derived from the initial data that is subjected to further analysis (Dülger et al. 2020: 8). However, this almost always occurs without the data subject's consent. Not to mention the possibility of the outputs breaching legal values that are within the scope of protection such as personal data (Abudureyimu and Oğurlu 2021: 773). In order to prevent this, the new/derived data must also be subjected to assessments within the scope of personal data protection (Abudureyimu and Oğurlu 2021: 766).
c. Personal data may be processed with a variety of data processing purposes that were not known from the beginning and are different from the original data collection purposes. Due to AI being unpredictable, reconstruction of the data processing purposes or relying on too many purposes to the point that it is no

2. is authorized by Union or Member State law to which the controller is subject, and which also lays down suitable measures to safeguard the data subject's rights and freedoms and legitimate interests; or
3. is based on the data subject's explicit consent".

longer compliant with the principle of being limited to processing purposes, may not be preventable (Dülger et al. 2020: 8).

8.2.4.3 Possible Solutions

In order to prevent violations of personal data protection principles and regulations within the scope of AI processes, various precautions including but not limited to the ones listed below must be taken:

- Taking into account fundamental human rights and the right to protection of personal data in AI processes (Dülger et al. 2020: 8).
- Educating and raising awareness in humans who are involved in the automated decision-making processes.
- Conducting risk assessments[9] in a sensitive manner that takes into account AI and big data's characteristics (Dülger et al. 2020: 9).
- Ensuring that the principle of data-protection-by-design (Dülger et al. 2020: 9; Turkish Personal Data Protection Authority 2021: 11) is complied with during automated decision-making processes.
- Carrying out data processing activities in accordance with the data minimization principle.
- Going through constant and periodic inspections and obtaining a self-regulatory approach.

To sum up, personal data protection and AI technologies are and always will be closely linked. In order to ensure the security and safety of the personal data that is fed to AI technologies, extensive measures must be taken. Otherwise, data breaches that will hinder, if not completely obliterate the benefits of using AI technologies are inevitable.

[9] Article 35 of GDPR titled "Data Protection Impact Assessment" regulates that; "Where a type of processing in particular using new technologies, and taking into account the nature, scope, context and purposes of the processing, is likely to result in a high risk to the rights and freedoms of natural persons, the controller shall, prior to the processing, carry out **an assessment of the impact of the envisaged processing operations on the protection of personal data**". For more information regarding Data Protection Impact Assessment, please see.

Information Commissioner's Office (ICO) "**DPIA**". Also, in the European Parliament's Resolution of 6 October 2021 on Artificial Intelligence in Criminal Law and Its Use By the Police and Judicial Authorities in Criminal Matters, the term "Fundamental Rights Impact Assessment" is used to describe a similar process. For more information, please see European Parliament (2021, par. Q. 20).

8.3 Cyber-Attacks

As it is general knowledge, no machine, system, program or software is one hundred percent immune to cyber-attacks. Whereas technology, and in parallel, means of obstructing that technology continuously advance, cybersecurity[10] is becoming more of an issue; with data leaks, data security breaches and unauthorized access to personal data and other information related to personal data continue to pose a significant threat (European Parliament 2021, see par. Q. 11) for the AI technologies, the personal data that is used in their decision-making processes and the other related information.

While it can make the process practical and fast, channeling an astounding amount of data onto one system can also cause the entire system to be at the risk of shutting down by one cyber-attack that incapacitates the entire system. This also jeopardizes the security and privacy of the personal data that is being used in the decision-making process creating further problems.

Even though it is absolutely impossible to be able to prevent cyber-attacks altogether, showcasing extreme diligence in ensuring cybersecurity, preferably in accordance with the security-by-design principle and upon specific human oversight (European Parliament 2021: par. Q. 11) is a sure way to minimize the risk to the possible extent.

8.4 Technical Challenges

Although it is not a very discussed pitfall of implementing AI to legal processes, current and possible technical difficulties should not be left out of the discussion since a theoretical idea may turn into an impossible task if the physical and technological means of the criminal justice and law enforcement practitioners are limited and the very advanced AI technology simply cannot be operated on. For example, as long as you do not have the infrastructure, the hardware, the computers, the wi-fi etc. to materialize your ideas, it is not possible to make them a reality, the good ideas just stay ideas.

At this point, it is necessary to point out that this issue is not specific to criminal justice or law enforcement. Technical challenges are known to present themselves in almost all areas AI systems are implemented to. However, we would be remiss to not mention this possible challenge that may adversely affect the implementation of AI to the legal profession on a large scale.

The efficiency of an AI technology lives and dies with its technical features and how well these technical features work. This fact may lead to a significant imbalance between different institutions or regions, even if they use the same AI technologies.

All regions of a country, or maybe even a city, may not be equally developed when it comes to technology. Hence, deciding to adapt an entire country to advanced, server

[10] For more information on cybersecurity in Turkey, please see Yazıcıoğlu et al. (2022a, b).

and internet consuming technologies that require at least a moderate level of expertise to use efficiently may bring legal processes to a halt in certain technologically underdeveloped parts of countries. This apparent and not-so-easy-to-solve asymmetry between technical capabilities (internet connection, number of servers, computers etc.) of nations or different regions of a nation, in large scale implementations, may severely hurt the overall performance of the AI technology.

With that being said, the possibility of technical difficulties should not cause the AI implementers to refrain from conducting large-scale implementations. In this sense, adopting a practical approach and focusing on the logistics of the aforementioned implementation is of great significance.

The principal solution that comes to mind for eliminating technical challenges would be conducting infrastructure strengthening in a large scale and making investments toward the necessary technological advancements (Dülger et al. 2020: 9). In this vein, it is of the utmost importance to work toward fixing the "asymmetry" between different regions where the same AI technology is or will be implemented. Assigning pilot cities or regions to firstly try out the AI technologies and avoiding picking these areas from only the technologically advanced ones would be optimal for overseeing the field conditions and fixing certain issues before it is too late.

Lastly, it must be pointed out that lack of technical prowess inevitably leads to lack of technical knowledge. This would pose a problem, especially in the older generations, who are known to be digital immigrants. Promoting and expediting nationwide and global AI literacy should be carried out swiftly and in parallel with improving technical conditions, since the people who will directly or indirectly take part in developing or implementing AI technologies need to gain the necessary knowledge and understanding with regard to how AI technology's function and what their effects on human rights are (Council of Europe Commissioner for Human Rights 2019: 14).

8.5 Conclusion

In this study, six of the biggest issues in usage of AI technologies in criminal justice system and law enforcement have been discussed.

The issues that were pointed out in this study and many more must thoroughly be examined and discussed in order to prevent or mitigate any damage or harm AI technologies may cause. This obviously calls for numerous efforts including but not limited to being mindful of biases that humans and AI technologies might have, making efforts in detecting and eliminating elements that cause discriminatory and/or biased decisions, protecting those who were and are being harmed by discriminatory and/or biased decisions as well as groups who are more likely to be discriminated against, evaluating any and all AI decision-making process within the scope of fundamental human rights, legal principles and protection of personal data principles, taking into account the presumption of innocence principle, especially in preventive policing activities, providing oversight on AI decision-making processes, taking steps toward ensuring transparency and accountability, making sure that AI technologies are secure and safe from cyber-attacks, promoting AI literacy and in the meantime, and supporting these actions with regulatory efforts.

We would like to emphasize that the existence and severity of these issues do not and should not overshadow the developments and improvements that were made possible thanks to AI technologies. When rightfully and lawfully used, AI technologies are extremely useful and efficient for both individuals and societies and they will continue to do so in the future. However, disregarding the obvious current and possible challenges and issues and not working toward preventing or mitigating the harm and damage that were, are, and will occur due to these issues would do nothing but steer us further from what we want to achieve.

References

Abudureyimu Y, Oğurlu Y (2021) Yapay Zekâ Uygulamalarının Kişisel Verilerin Korunmasına Dair Doğurabileceği Sorunlar ve Çözüm Önerileri. İstanbul Ticaret Üniversitesi Sosyal Bilimler Dergisi 20(41):765–782. https://doi.org/10.46928/iticusbe.863505

American Psychological Association (APA) Dictionary. https://dictionary.apa.org/anchoring-bias (Last Accessed: 10 May 2022)

Bathaee Y (2018) The artificial intelligence black box and the failure of intent and causation. Harv J Law Technol 31(2) Spring 2018. https://jolt.law.harvard.edu/assets/articlePDFs/v31/The-Artificial-Intelligence-Black-Box-and-the-Failure-of-Intent-and-Causation-Yavar-Bathaee.pdf (Last Accessed: 10 May 2022)

Borgesius FZ (2018) Discrimination, artificial intelligence and algorithmic decision-making. Council of Europe Directorate General of Democracy, Strasbourg. https://rm.coe.int/discrimination-artificial-intelligence-and-algorithmic-decision-making/1680925d73 (Last Accessed: 10 May 2022)

Brittanica. https://www.britannica.com/science/confirmation-bias (Last Accessed: 10 May 2022)

Büyüksağiş E (2021) Yapay Zeka Karşısında Kişisel Verilerin Korunması ve Revizyon İhtiyacı. YÜHFD C.XVIII, 2021/2, pp 529–541

Cambridge Dictionary. https://dictionary.cambridge.org/us/dictionary/english/cognitive-bias (Last Accessed: 10 May 2022)

Council of Europe Commissioner for Human Rights (2019) Unboxing artificial intelligence: 10 steps to protect human rights. Council of Europe. https://rm.coe.int/unboxing-artificial-intelligence-10-steps-to-protect-human-rights-reco/1680946e64 (Last Accessed: 10 May 2022)

Council of Europe Directorate General of Human Rights and Rule of Law (2019) Consultative committee of the convention for the protection of individuals with regard to automatic processing of personal data (convention 108)—Guidelines on artificial intelligence and data protection, T-PD (2019)01, Strasbourg. https://rm.coe.int/guidelines-on-artificial-intelligence-and-data-protection/168091f9d8 (Last Accessed: 10 May 2022)

Demirtaş ZC (2021) İstanbul bar association IT law commission artificial intelligence study group Bulletin - Yapay Zeka Çağında Hukuk, 13th Issue, İstanbul. https://www.istanbulbarosu.org.tr/files/komisyonlar/yzcg/2021ekimbulten.pdf (Last Accessed: 10 May 2022)

Dülger MV, Çetin S, Aydınlı CD (2020) Görüş: Algoritmik Karar Verme ve Veri Koruması. İstanbul Bar Association IT Law Commission Artificial Intelligence Study Group, İstanbul. https://www.istanbulbarosu.org.tr/files/docs/AlgoritmikKararVermeveVeriKorumas%C4%B1022020.pdf (Last Accessed: 10 May 2022)

European Parliament (2021) Resolution of 6 October 2021 on artificial intelligence in criminal law and its use by the police and judicial authorities in criminal matters, (2020/2016(INI)),

Strasbourg. https://www.europarl.europa.eu/doceo/document/TA-9-2021-0405_EN.html (Last Accessed: 10 May 2022)

Fair Trials (2021) Automating injustice: the use of artificial intelligence & automated decision-making systems in criminal justice in Europe. https://www.fairtrials.org/app/uploads/2021/11/Automating_Injustice.pdf (Last Accessed: 10 May 2022)

Information Commissioner's Office (ICO). Data Protection Impact Assessments ("DPIA"). https://ico.org.uk/for-organisations/guide-to-data-protection/guide-to-the-general-data-protection-regulation-gdpr/accountability-and-governance/data-protection-impact-assessments/ (Last Accessed: 10 May 2022)

Information Commissioner's Office (ICO). Principle (c): Data Minimisation ("Data Minimisation"). https://ico.org.uk/for-organisations/guide-to-data-protection/guide-to-the-general-data-protection-regulation-gdpr/principles/data-minimisation/ (Last Accessed: 10 May 2022)

İçer Z (2021) Yapay Zekâ Temelli Önleyici Hukuk Mekanizmaları - Öngörücü Polislik, İstanbul bar association IT law commission artificial intelligence study group annual report: artificial intelligence based technologies and criminal law, İstanbul, pp 29–47

Schwartz R, Vassilev A, Greene K, Perine L, Burt A, Hall P (2022) Towards a standard for identifying and managing bias in artificial intelligence, NIST (National Institute of Standards and Technology) Special Publication 1270. https://doi.org/10.6028/NIST.SP.1270

Turkish Personal Data Protection Authority (2021) Recommendations regarding protection of personal data in the field of artificial intelligence. https://www.kvkk.gov.tr/Icerik/7048/Yapay-Zeka-Alaninda-Kisisel-Verilerin-Korunmasina-Dair-Tavsiyeler (Last Accessed: 10 May 2022)

Yazıcıoğlu B, İslamoğlu Bayer K, Çaylak B, İnan VM (2022a) Cybersecurity 2022: Turkey practice guide. Chambers and Partners. https://practiceguides.chambers.com/practice-guides/data-protection-privacy-2022/turkey (Last Accessed: 10 May 2022)

Yazıcıoğlu B, İslamoğlu Bayer K, Özcanlı İİ, Uygun E (2022b) Data protection & privacy 2022: Turkey practice guide. Chambers and Partners. https://practiceguides.chambers.com/practice-guides/data-protection-privacy-2022/turkey (Last Accessed: 10 May 2022)

Benay Çaylak She graduated from Istanbul University Faculty of Law in 2017 and Anadolu University Web Design and Coding Department in 2020. Currently, she is continuing her associate degree education at Anadolu University Health Management and Istanbul University Management Information Systems, and her master's degree in Private Law at Istanbul University Social Sciences Institute. She is a member of Istanbul Bar Association Medical Law Center, Istanbul Bar Association Information Law Commission and Personal Data Protection Commission. She is currently working as a lawyer at Yazıcıoğlu Law Firm, affiliated with the Istanbul Bar Association.

Part VI
Evaluation of the Interaction of Law and Artificial Intelligence Within Different Application Areas

Chapter 9
Artificial Intelligence and Prohibition of Discrimination from the Perspective of Private Law

Ş. Barış Özçelik

Abstract Artificial intelligence (AI) technologies promise to change our lives in positive way in many aspects while bringing along some risks. One of these risks is the possibility that decisions based on AI systems contain discrimination. Since the prohibition of discrimination is predominantly seen as a matter of public law, it may seem to be questionable to talk about the prohibition of discrimination in private law where principles of private autonomy and particularly freedom of contract prevail. Nevertheless, depriving individuals of the opportunity to enter into a fair and freely negotiated contract as a result of discrimination would be incompatible with the ideas underlying the freedom of contract. Moreover, since discrimination is insulting in most of the cases, it also violates the personal rights of the individual who is discriminated against. Thus, discrimination is an issue that also needs to be considered from the perspective of private law. As private law sanctions, nullity, compensation or an obligation to contract can be applied against discrimination. The fact that discrimination is the product of a decision-making mechanism using AI systems brings along some legal problems specific to this situation. One of these problems is that the results produced by some AI technologies are unexplainable since the reasons on which the decision is based must first be known to conclude that a decision is based on discrimination.

Keywords Artificial intelligence · Discrimination · Private law · Freedom of contract · Explainability

JEL Codes K38 · K39 · K31 · K13 · K12

Ş. B. Özçelik (✉)
Faculty of Law, Bilkent University, Ankara, Turkey
e-mail: bozcelik@bilkent.edu.tr

M. Kılıç and S. Bozkuş Kahyaoğlu (eds.), *Algorithmic Discrimination and Ethical Perspective of Artificial Intelligence*, Accounting, Finance, Sustainability, Governance & Fraud: Theory and Application, https://doi.org/10.1007/978-981-99-6327-0_9

9.1 Introduction

Discrimination is one of the risks posed by the spread of the use of artificial intelligence in daily life. This risk, which mainly arises from the data-drivenness of AI systems, exists for decisions taken as to certain activities such as recruitment, education, insurance, credit scoring, or marketing.

The Constitution of the Republic of Türkiye of 1982 stipulates the principle of equality and the prohibition of discrimination. The European Convention on Human Rights, which includes the prohibition of discrimination and is a part of domestic law, has been in force in Türkiye since 18 May 1954.

The Law on Human Rights and Equality Institution of Türkiye, which came into force on April 20, 2016, includes the definition of discrimination, the scope of the prohibition of discrimination in terms of subject and person, grounds of discrimination, and administrative sanctions against discrimination. In terms of person and subject, the mentioned Law covers both the natural and legal persons of private law and the private law relations within the scope of the prohibition of discrimination. However, the Law does not include any provisions on private law sanctions that can be applied against discrimination.

It has long been accepted that the prohibition of discrimination also applies in private law. Because discrimination is not only contrary to the fundamental values of legal systems that put human dignity at the center but also incompatible with the underlying ideas of private law. Indeed, the Law on Human Rights and Equality Institution of Türkiye has adopted this approach by including both private law persons and private law relations in its scope of application. However, it is necessary to clarify which private law sanctions can be applied under what conditions and in what scope against discrimination, in addition to the administrative sanctions foreseen by the aforementioned Law.

On the other hand, it is known that today, AI is used as a support system for decisions about individuals in many sectors and there is a possibility that these decisions may be discriminatory because of the data-drivenness of AI systems. The involvement of AI can create some problems that affect the detection of discrimination and the implementation of sanctions in terms of private law. In this regard, the fact that some AI systems are not explainable seems to be the biggest problem.

Against this background, this paper aims to analyze the application of the prohibition of discrimination in private law relations, private law sanctions that may be applied against discrimination, and the legal effects of the involvement of AI in these respects.

9.2 Defining Discrimination

Paragraph 1 of Article 10 of the Constitution of the Republic of Türkiye also provides that "Everyone is equal before the law without distinction as to language, race, colour, sex, political opinion, philosophical belief, religion, and sect, or any such grounds".

European Convention on Human Rights, which is a part of domestic law, provides under Article 14 that "The enjoyment of the rights and freedoms set forth in this Convention shall be secured without discrimination on any ground such as sex, race, colour, language, religion, political or other opinion, national or social origin, association with a national minority, property, birth or other status".

According to Article 3 of the Law on Human Rights and Equality Institution of Türkiye, "Everyone is equal in enjoying legally recognized rights and freedoms (Paragraph 1). Under this law, discrimination based on gender, race, color, language, religion, faith, sect, philosophical and political opinion, ethnicity, wealth, birth, marital status, health status, disability, and age is prohibited" (Paragraph 2).

As a legal term, discrimination can be defined in several ways, depending on the aim and scope of the related legal instrument. Furthermore, some distinctions are made such as direct and indirect discrimination.

Cambridge Dictionary defines discrimination generally as "treating a person or particular group of people differently, especially in a worse way from the way in which you treat other people, because of their skin colour, sex, sexuality, etc." (Cambridge Dictionary 2022).

According to Amnesty International, "Discrimination occurs when a person is unable to enjoy his or her human rights or other legal rights on an equal basis with others because of an unjustified distinction made in policy, law or treatment" (Amnesty International 2022).

Although the definition of discrimination is not directly within the scope of this paper, for its purposes, the term discrimination refers to the negative difference of any treatment applied to a natural or legal person from that applied to other persons who have the same conditions, without any objective and reasonable ground (For a similar definition see, European Court of Human Rights (2007)).

9.3 Private Law and Prohibition of Discrimination

Prohibition of discrimination, which primarily addresses government bodies and takes its origin from constitutions, international treaties, and other related legislation, is predominantly seen as a matter of public law. For example, Paragraph 5 of Article 10 of the Constitution of the Republic of Türkiye, provides that "*State organs and administrative authorities* are obliged to act in compliance with the principle of equality before the law in all their proceedings".

In private law, on the other hand, the principle of private (party) autonomy prevails. Many social and economic liberties including freedoms of contract, testament, association, work, enterprise, and even private property rely on said principle. The most basic function of private law is to give individuals the opportunity to create their own legal relationships under their own responsibility. Relying on the assumption that individuals are equal, private law reflects the idea of commutative justice (Looschelders 2012).

Thus, it may seem to be questionable to talk about the prohibition of discrimination, which considers the special conditions of individuals and cases and reflects the idea of distributive justice (Looschelders 2012), in private law. The fact that private law gives individuals the freedom to contract with anyone and under any conditions they want or even not to contract at all, may justify such a doubt, at least at first sight.

However, when one takes a closer look at the subject, it becomes clear that the relationship between private law and the prohibition of discrimination is different from what it seems at first glance:

First of all, the prohibition of discrimination should be seen as one of the fundamental values of any legal system that puts human dignity at a central place and adopts the principle of equality. As a matter of fact, besides the Constitution, regulations including the European Convention on Human Rights (which is a part of domestic law), Law on Human Rights and Equality Institution of Türkiye, and the Turkish Penal Code (Article 122) support the idea that prohibition of discrimination should be seen as one of the fundamental values of Turkish legal system. Therefore, it also includes private law relations. European Court of Justice has also confirmed this approach from the standpoint of European Union Law in its *Test-Achats* judgement of 1 March 2011 (European Court of Justice 2011) (for an assessment of mentioned decision, see, Reich (2011)).

Secondly, according to the Article 2 of the Turkish Civil Code, which is the positive basis of the principle of good faith in Turkish law, everyone must comply with the rules of honesty when exercising their rights and fulfilling their obligations and the law does not protect the blatant abuse of any right. The refusal of a contractual offer for discriminatory reasons means that the refuser does not have any legitimate interest in avoiding contracting and abuses the freedom of contract. Therefore, such a discriminatory refusal can not be legally protected, since it constitutes a violation of the principle of good faith, which applies to all private legal relationships (Demirsatan 2022).

On the other hand, discrimination is insulting in most of the cases, and it violates the personal rights of the individual, which are under the protection of Article 24 of the Turkish Civil Code. In other words, freedom of contract does not give the right anyone to violate the personal rights of others.

To oppose the application of the prohibition of discrimination in private law relations relying on the freedom of contract would mean understanding the said freedom only in its formal sense and ignoring its essence. In fact, discrimination is incompatible with the ideas underlying the freedom of contract, including to maintain a free and competitive market, since it means depriving individuals of the opportunity to enter into a fair and freely negotiated contract (Looschelders 2012). The function of the prohibition of discrimination here can be compared to provisions that protect disadvantaged parties such as consumers, workers, and renters.

The provisions of the Turkish Civil Code regarding the capacity to have rights and the capacity to act also confirm the idea that discrimination is incompatible with the ideas underlying private law. Article 8 of the Code states that all people are equal in their capacity to have rights and obligations within the boundaries of the law, while Article 9 states that anyone who has the capacity to act can acquire rights and undertake obligations with their own acts. Therefore, it is not possible to accept that, discriminatory treatments, which means that some individuals or groups are excluded from or disadvantaged in economic life, are compatible with the ideas underlying the mentioned provisions.

All these facts make it clear that the prohibition of discrimination should also be taken into account in terms of private law. Indeed, Article 5 of the Law on Human Rights and Equality Institution of Türkiye, which regulates the scope of the prohibition of discrimination, confirms and clarifies this approach by including many private law relations.

9.4 Private Law Sanctions Against Discrimination

After determining that the prohibition of discrimination involves also in private law relations, it is necessary to determine what would be sanctions against discrimination in terms of private law. While providing for the administrative fine against discriminatory treatment, The Law on Human Rights and Equality Institution of Türkiye does not include any provisions on private law sanctions. Therefore, private law sanctions should be determined according to the general provisions.

9.4.1 Nullity

One of the conceivable private law sanctions is the nullity, which is applied to unlawful legal acts in general. Article 27 paragraph 1 of the Turkish Code of Obligations stipulates those contracts which violate mandatory provisions of the law, morality, public order, or personal rights or contracts, subject of which are impossible, are null and void. In this context, contracts that contain discriminatory provisions must also be held as null and void, since, as stated above, it contradicts one of the fundamental values, namely prohibition of discrimination, of the Turkish legal system.

However, since, according to the second paragraph of the mentioned provision, the fact that some of the provisions contained in the contract are null and void does not affect the validity of the other provisions, nullity sanction shall be applied partially, i.e. the contract shall be deemed to be concluded without discriminatory provisions. Moreover, considering the essence of the prohibition of discrimination, the discriminating party can not claim that the contract should completely be held as null and void, arguing that she/he would not have concluded the contract at all without null and void provisions. For example, in an employment contract, if an employee is disadvantaged compared to another employee under conditions that would constitute discrimination, the contract shall be deemed to have been concluded without such conditions.

On the other hand, since the prohibition of discrimination also applies to all legal acts, for example, a discriminatory provision in the statute of an association or in a testament is also held as null and void (European Court of Human Rights 2019).

9.4.2 Compensation

Another applicable private law sanction against discrimination is compensation. The person who is discriminated against may seek compensation for material and immaterial damages she or he has suffered because of the discriminatory treatment. The basis and amount of possible compensation change depending on the circumstances of the specific case. For example, if discrimination occurs during contract negotiations, the basis of possible compensation would be *culpa in contrahendo* (Türkmen 2017). If conditions of the violation of personal rights are met in the same case, compensation for immaterial damages can also be claimed under Article 58 of the Turkish Code of Obligations. If discrimination occurs during the performance of an existing contract, on the other hand, the person who is discriminated against may claim compensation under the provisions on breach of contract, under Article 112 ff. of the Turkish Code of Obligations.

The amount of compensation is also determined by considering the type of damage suffered in the specific case. For example, if a person, whose offer to contract is refused due to discriminatory reasons, has to enter into a contract in which she or he got the same goods or services for a higher price, the extra amount she or he paid constitutes her or his loss and thus, the amount of compensation. However, special provisions in this regard should also be taken into account. For example, according to Article 5 of the Labor Law, the worker who is discriminated against in labor relations may claim an appropriate compensation of up to 4 months' wages, besides the rights of which she or he is deprived.

9.4.3 Obligation to Contract

Accepting that the discriminating party has an obligation to contract also can be seen as a sanction against discrimination. In some cases, this may be an appropriate sanction against discrimination since the person who is discriminated against can not be expected to put up with it (Looschelders 2012). Undoubtedly, the basic requirement of the existence of an obligation to contract is that the contract could be concluded if there had not been discrimination. Under this condition, the obligation to contract can generally be based on the principle of good faith in terms of Article 2 of the Turkish Civil Code (For other possible bases of obligation to contract see, Türkmen (2017)).

Accordingly, the person whose offer to contract is rejected for discriminatory reasons can sue to ensure the specific performance of the other party's obligation to accept. When the court accepts the case, the rejected contract is established. The person who is discriminated against can also assert his or her rights arising from the rejected contract within the same case (Demirsatan 2022).

9.4.4 Burden of Proof

One of the most important challenges that can be faced in the application of private law sanctions against discrimination is the burden of proof. As a rule, since the person who is discriminated against will have to prove it, many claims based on this may fail, particularly in cases where the fact that constitutes discrimination is not expressly declared. Considering this fact, Article 21 of The Law on Human Rights and Equality Institution of Türkiye states that if the applicant demonstrates the existence of strong indications of the reality of his claim and the facts that constitute a presumption of discrimination, the other party must prove that she or he did not violate the prohibition of discrimination and the principle of equal treatment. Paragraph 8 of Article 5 of the Labor Law also states that when the employee demonstrates a fact that strongly indicates the possibility of the existence of a violation of the prohibition of discrimination, the employer has to prove that such a violation does not exist.

Thus, under the circumstances that are specified in the above-mentioned provisions, the burden of proof is reversed and the other party has to prove that it did not discriminate. Considering the difficulty in proving discrimination, the underlying idea of these provisions should be accepted as a general principle in terms of any claims based on discrimination, including private law ones. This has special importance in terms of the use of artificial intelligence, where technological complexity is also involved.

9.5 Artificial Intelligence and Discrimination

As is known, one of the most significant features of AI is data-drivenness. In order to understand the impact of this feature of AI on its potential to produce discriminatory decisions, it is necessary to explain briefly the decision-making process of AI: In order to achieve a certain decision through an AI system, a large amount of data are needed (big data). Some of this large amount of data (training data) are divided into subsets by human users (labeling) according to the output targeted. What is expected from the system is to find relations (correlations) between elements contained in the subsets, which are consistent with the targeted result. When an AI system discovers those correlations, it "learns" the criteria necessary to achieve the targeted result and uses them to produce related decisions (for a detailed explanation of the process see, Barocas and Selbst (2016)).

As this brief explanation demonstrates, it is possible that the data set used contains discriminatory data or that discriminatory criterion is used in the determination of mentioned subsets, and even that the intended result is determined according to the discriminative criteria in the decision-making process of AI. In all these cases, it is inevitable that the decisions produced by AI will be discriminatory.

As a matter of fact, some well-known real-life examples confirm this finding. An AI-based chatbot, TAY, released by Microsoft via Twitter, soon began posting racist and sexist messages. This was explained by the fact that the data set that consisted of other users' tweets and from which TAY learned was biased (Reuters 2016). Likewise, in the United States, where the risk of re-committing a crime is taken into account in determining the decision about offenders, it was observed that the AI-based program COMPAS, which is used to determine the risk in question, discriminates between white and black people and calculate the risk for latter higher than the former. It was claimed that this was because of the fact that the data set used to train the algorithm COMPAS was discriminatory against black people (Angwin et al. 2016). Finally, an AI-based recruitment program used by Amazon was reported to be discriminatory against female job applicants for certain posts, again because of the biased training data (Reuters 2018) (for more examples see, Borgesius (2018)).

9.6 AI-Induced Discrimination and Its Consequences in Terms of Private Law

As is known, AI systems are used as decision support systems in many sectors. Thus, a natural or legal person makes a decision relying on the results produced by the AI system. Considering the fields where AI systems are widely used, it is possible to foresee that, discriminatory decisions or treatments that rely on the use of artificial intelligence can be faced in matters concerning private law, for example, in deciding on recruitment or employment conditions, determining the credit score of individuals or price of goods or services or on insurance issues.

9.6.1 The Black-Box Problem

The first problem that can be faced in this regard is the determination of the existence of discrimination. As mentioned above, this difficulty, which already exists in terms of discrimination even in cases where AI is not involved, has special importance regarding decisions made based on AI systems, where technological complexity involves. This arises from the fact that the results produced by AI systems are unexplainable, particularly the ones that rely on AI systems based on deep learning techniques. In this regard, being unexplainable that is also called as the black-box problem, which means that the reasons for a decision produced by the AI system can not be understood by human users. Since the reasons for the result produced by the AI system are not known, it is difficult to determine whether the decision made is discriminatory.

Against this problem, the following solution can be proposed: In cases where a decision made or treatment applied regarding a person based on an AI system and the reasons for the decision or treatment applied to that person can not be explained, the burden of proving that the relevant decision or treatment is not based on discrimination should be placed on the party that implements the relevant decision or treatment. This party may fulfill mentioned burden by proving its legitimate interest in the relevant decision or application. The solution proposed here is nothing more than the adaptation of the already mentioned provisions of The Law on Human Rights and Equality Institution of Türkiye (Article 21) and the Labor Law (Article 5, paragraph 8), which provide for the reversal of the burden of proof for the claims arising from discrimination under certain circumstances, to the cases in which AI is involved.

9.6.2 Sanctions Against AI-Induced Discrimination

The second point to clarify is what are the private law sanctions that can be applied against AI-induced discrimination. Undoubtedly, the answer to this question will change depending on the way discrimination occurs.

For example, if the contractual provisions prepared by an AI program are discriminatory, nullity sanction of private law may apply against the provisions of such a contract.

Again, for example, if a person's job application or offer to enter into an insurance contract is rejected solely due to AI-induced discrimination that person may apply to the court and claim the specific performance of the obligation to accept based on the principle of good faith and thus the conclusion of the contract. In the same examples, besides the conclusion of the contract, compensation for the damage suffered due to the rejection of the contract offer (damages for delay of the acceptance) may also be claimed.

Likewise, since no one can be forced to contract with a person who discriminates against him or her, the person who is discriminated against may also seek compensation for the damage he or she suffered due to non-conclusion of the contract, rather than claiming the establishment of the contract. For example, a person may have concluded another employment contract for a lower wage or insurance contract for a higher price and claim the differences as material damages. Similarly, anyone who pays for a good or service more than others because of discriminatory reasons (price discrimination) can ask for a refund of the extra part of the price she or he paid.

In all the above-mentioned examples, depending on the existence of the conditions of the violation of the personal rights, the person who is discriminated against may also seek compensation for immaterial harm she or he suffered, for instance, due to sadness, stress, or loss of reputation.

As a matter of course, in all cases, causality between discrimination and harm or damage should be proven by the person who is discriminated against.

9.6.3 Who is Liable for AI-Induced Discrimination?

Another question that needs to be clarified in terms of AI-caused discrimination is who will be sanctioned against discrimination, in other words, who will be held liable for the consequences of AI-induced discrimination in terms of private law. In fact, there is no doubt that the natural or legal person who makes the final discriminatory decision is liable for the damage that occurred because of the decision. This natural or legal person can not escape from liability by arguing that discriminatory decision relies on the results produced by the AI system. Otherwise, for example, if the same results were obtained from the activity of an employee, the person who would be liable based on vicarious liability would be exempted from that, just because she or he used the AI system instead of the work of an employee. It is clear that this would not be a fair result.

Since the discrimination and harm or damage arising from that is the result of the final decision which is made by the person who uses the AI system, according to the existing liability rules, those who develop and/or update the AI system can not directly be held liable against the victim of discrimination. In other words, in cases where the AI system is used to support human decisions, it can not be accepted that there is causality between the defect of the AI system and harm or damage that occurred. Thus, the liability of developers, etc. can only be at stake in the recourse action against them by the person who used the AI system and paid compensation for discrimination-caused harm or damage.

This issue concerns liability for AI-related damages in general, regardless of discrimination. It is suggested that those who develop and update AI systems should be accountable besides those who use AI systems, as the latter's control over system-induced risks is very limited (European Commission, NTF Report 2019). Following this approach, in the European Parliament's Resolution titled "Civil Liability Regime for Artificial Intelligence" of 20 October 2020 and in the Proposal for a Regulation attached to mentioned Resolution, it is suggested that the operators of such systems should be held liable for damages caused by AI. The term operator in this sense includes both the person who uses the system (frontend operator) and the person who determines the characteristics of the technology and provides data and technical support service in the background in a continuous manner (backend operator). According to the proposal, these two types of operators are jointly and severally liable before the persons who suffer damages (for a critical analysis of the mentioned proposal see, Özçelik (2021)).

9.7 Conclusion

Artificial intelligence promises to make our lives easier in many aspects, while also bringing along some risks including discrimination, because of its data-drivenness.

As fundamental values of the legal system, principles of equality and prohibition of discrimination also prevail in private law.

Depending on the circumstances of the specific case, nullity, compensation, or an obligation to contract can be at stake as private law sanctions against AI-caused discrimination. Compensation, in this sense, includes both material and immaterial harm or damage suffered, provided that causality between discrimination and harm or damage is proved.

The most important obstacle in terms of the application of private law sanctions against discrimination is that the results produced by some AI systems are unexplainable. Since the reasons for the result produced by the AI system are not known, it is difficult to determine whether the decision made is discriminatory. To solve this problem, the burden of proving that the relevant decision or treatment is not based on discrimination should be placed on the party that implements the relevant decision or treatment.

The natural or legal person who makes the final discriminatory decision upon results produced by the AI system is liable for the consequences of AI-induced discrimination in terms of private law. This person can not escape from liability by arguing that discriminatory decision relies on the results produced by the AI system. According to the existing liability rules, those who develop and update AI system can not directly be held liable against the victim of discrimination. On this issue, the general approach, which will be adopted by lawmakers in the future, concerning liability for the damages caused by AI will be decisive.

References

Amnesty International (2022) Discrimination. https://www.amnesty.org/en/what-we-do/discrimination/. Accessed 15 Apr 2022

Angwin J, Larson J, Mattu S, Kirchner L (2016) Machine bias: there's software used across the country to predict future criminals. And it's biased against blacks, ProPublica. https://www.propublica.org/Article/machine-bias-risk-assessments-in-criminal-sentencing. Accessed 15 Apr 2022

Barocas S, Selbst AD (2016) Big data's disparate impact. Calif Law Rev 104:671–732

Borgesius FZ (2018) Discrimination, artificial intelligence and algorithmic decision-making. Council of Europe, Directorate General of Democracy, Strasbourg

Cambridge Dictionary (2022) Discrimination. https://dictionary.cambridge.org/tr/s%C3%B6zl%C3%BCk/ingilizce/discrimination. Accessed 15 Apr 2022

Demirsatan B (2022) Türk Borçlar Hukukunda Sözleşme Yapma Zorunluluğu, İstanbul, Filiz Kitabevi

European Commission (2019) Report on liability for AI and other digital technologies-new technologies formation, (NTF Report). https://ec.europa.eu/transparency/regexpert/index.cfm?do=groupDetail.groupMeetingDoc&docid=36608. Accessed 15 Apr 2022

European Court of Human Rights (2007) D.H. and others v. The Czech Republic, Application no. 57325/00

European Court of Human Rights (2019) Deaconu and Alexandru Bogdan v. Romania, Application no. 66299/12

European Court of Justice (2011) Association belge des Consommateurs Test-Achats ASBL and Others v. Conseil des ministers, Case C-236/09

Looschelders D (2012) Diskriminierung und Schutz vor Diskriminierung im Privatrecht. Juristen Zeitung 67(3):105–114

Özçelik ŞB (2021) Civil liability regime for artificial intelligence, a critical analysis European parliament's proposal for a regulation. Eur Leg Forum 21(5–6):93–100

Reich N (2011) Non-discrimination and the many faces of private law in the union–some thoughts after the "Test-Achats" judgment. Eur J Risk Regul 2(2):283–290

Reuters (2018) Amazon scraps secret AI recruiting tool that showed bias against women. https://www.reuters.com/Article/us-amazon-com-jobs-automation-insight-idUSKCN1MK08G. Accessed 15 Apr 2022

Reuters (2016) Microsoft's AI Twitter bot goes dark after racist, sexist tweets. https://www.reuters.com/Article/us-microsoft-twitter-bot-idUSKCN0WQ2LA. Accessed 15 Apr 2022

Türkmen A (2017) 6701 Sayılı Kanunda Yer Alan Ayrımcılık Yasağının Sözleşme Hukukuna Etkilerine İlişkin Genel Bir Değerlendirme. Bahçeşehir Üniversitesi Hukuk Fakültesi Dergisi 12(149–150):135–178

Ş. Barış Özçelik A faculty member of Bilkent University Faculty of Law, graduated from Ankara University Faculty of Law and received the titles of "doctor" in the field of civil law in 2009 and "associate professor" in the same field in 2018. In 2007–2008, he received the Swiss Federal Government Scholarship and continued his Ph.D. on "force majeure in contract law" at University of Basel. In 2019, he was awarded TUBITAK support with his two years research project on "Artificial Intelligence and Law". Dr. Özçelik has published numerous national and international scientific articles and a book on various subjects and speaks English and German.

Chapter 10
Legal Challenges of Artificial Intelligence in Healthcare

Merve Ayşegül Kulular İbrahim

Abstract Right to health is a fundamental right defined in World Health Organization 1946 as "the enjoyment of the highest attainable standard of health is one of the fundamental rights of every human being". Because right to health is recognized as human right, every human being must have access to health services without distinction of race, gender, or economic status. In addition, beyond health services, there are several rights such as the right to freedom from discrimination, and the right to benefit from scientific progress and its applications providing sufficient protection for right to health. Accordingly, to promote right to health information technologies play a significant role. Due to technologic developments, artificial intelligence (AI) is used in healthcare for a number of implementations. AI applications provide many advantages in medical care. However, there are also significant risks of AI applications. AI might cause infringement of the right to health considering discrimination. There are several research illustrating discrimination due to algorithmic bias in health sector. The discrimination might be unintentionally and more. AI systems are capable of learning progression. The data that AI systems are designed from might incorporate cognitive biases. AI system learns from both training data and its own experience. As a result, the AI learning progression may unintentionally lead more discrimination and hurt patients. To prevent algorithmic discrimination and infringement on human rights, this work proposes not only new laws and policies but also measures for standards of technical tools. Last, but not least considering that training data that causes decision-making bias of AI applications, is consist of health care professionals' decisions, health care professionals should be educated about antiracism to provide sufficient protection on right to health.

Keywords Artificial intelligence · IT law · Cyber law · Deep learning · Human rights · Prohibition of discrimination

JEL Codes K24 · K37 · K38

M. A. K. İbrahim (✉)
Social Science University of Ankara, Ankara, Turkey
e-mail: aysegul.kulular@asbu.edu.tr

M. Kılıç and S. Bozkuş Kahyaoğlu (eds.), *Algorithmic Discrimination and Ethical Perspective of Artificial Intelligence*, Accounting, Finance, Sustainability, Governance & Fraud: Theory and Application, https://doi.org/10.1007/978-981-99-6327-0_10

10.1 Introduction

Artificial intelligence is used in various fields. One of these areas is the health sector. In the health sector, artificial intelligence is used for purposes such as; increasing efficiency in medical research and quality as a result of feedback (Packin and Lev-Aretz 2018: 107), extracting of medical records (Ng et al. 2016: 650), guiding physicians in clinical decision-making in planning treatment (Deo and Nallamothu 2016: 618), accurate diagnosis of patients with the analysis of electronic health records and even the ability to diagnose priority conditions such as heart failure 2 years in advance (IBM Research Editorial Staff 2017), detection of diseases with better quality and faster imaging (Locklear 2017), the possibility of foreseeing the harm that may occur in the health of the patient 2 days before they occur (Suleyman and King 2019), estimating the waiting times of hospitalized patients, predicting the probability of recovery, reducing cost (Suleyman and King 2019). In general, the use of artificial intelligence in the health sector provides various conveniences related to the right to health, which is one of the most fundamental rights. Artificial intelligence has important effects on three different groups in the field of health, in terms of doctors, health systems, and patients (Topol 2019: 44). For example, detecting pathology or imaging organs using low radiation dose with collected data is possible through the artificial intelligence. With artificial intelligence in computed tomography (CT), both the radiation dose is reduced and the probability of error is reduced (McCollough and Leng 2020: 113; Topol 2019: 44). This shows the effect of artificial intelligence on doctors to make the correct diagnosis. With the effect of artificial intelligence on health systems, access to patient information from miles away is provided through the use of artificial intelligence in health technologies (Roberts and Luces 2003: 19). The effect of the use of artificial intelligence in health services on patients is that patients can process the data obtained by artificial intelligence from various sources such as electronic health records, medical literature, clinical trials, insurance data, pharmacy records and social media content of patients (Price II 2017: 10) (Topol 2019: 44). The diagnosis is made by taking into account the patient's history and then the treatment method is determined (Sarı 2020: 252) (Price II 2017: 10). By using artificial intelligence, the probability of error in the interpretation of the disease is reduced (McCollough and Leng 2020: 113) (Price II 2017: 10). The use of artificial intelligence in the health sector has both advantages and disadvantages. Practices that contradict the prohibition of discrimination are at the forefront of these harms Discrimination is prohibited in article 3 of the Turkish Human Rights and Equality Institution Law No. 6701. Under the title of Hate and Discrimination in Article 122 of the Turkish Penal Code No. 5237, discrimination is prohibited;

> Because of hatred based on language, race, nationality, color, gender, disability, political opinion, philosophical belief, religion or sect, anyone;
>
> (a) who prevents the sale, transfer or rental of a movable or immovable property that has been offered to the public,
>
> (b) who prevents a person from benefiting from a certain service offered to the public,
>
> (c) who prevents a person from being hired,

(d) who prevents a person from engaging in an ordinary economic activity
(e) is punished with imprisonment from one to three years.

Artificial intelligence can make algorithmic decisions that violate the prohibition of discrimination in the health sector (Cofone 2019: 1389). This is because of the data contained in the *training data*, which usually enables the artificial intelligence to make decisions. If discriminatory datasets are used to train algorithms, certain individuals or groups will be disfavored when making decisions (Criado and Such 2019: 8). Another reason why artificial intelligence makes discriminatory decisions is that the boundaries of the data in the training data are not certain. In other words, there are no effective regulations to train the algorithm with data that will not cause discrimination. Considering the benefits of algorithmic decision-making mechanisms created using artificial intelligence, within the scope of the prohibition of discrimination, decision-making mechanisms consisting of algorithms should be regulated legally, rather than prohibiting these algorithms (Cofone 2019: 1391).

This chapter focuses on the usage of artificial intelligence in healthcare regarding prioritization of image review, and prioritization of admission to hospital. Considering discrimination caused by artificial intelligence, it discusses liability of different actors such as artificial intelligence, healthcare professionals, the operator, and the manufacturer or software developer. It aims to demonstrate positive and negative impacts of artificial intelligence in decision-making concerning prohibition of discrimination among patients.

10.2 Situations Where Artificial Intelligence May Cause Discrimination Among Patients

a. **Prioritization of Image Review**

It is possible to obtain more accurate images with lower radiation in the use of information technologies in health services, especially in imaging devices. Also, it is possible to clarify the image or remove ambiguities by using artificial intelligence. Moreover, some companies can provide priority examination of the patient's image if artificial intelligence detects important diseases such as pneumothorax during imaging (Miliard, 2018). In case the disease diagnoses specified during data entry to the artificial intelligence are detected by artificial intelligence during imaging, the images of that patient will be evaluated by the doctor before other patients. However, here, there may be situations such as the radiologist doctor may not specify the patient as a priority or may not be able to detect the image, although the artificial intelligence detects the image, and although the artificial intelligence specifies the patient as a priority, the radiologist doctor reports the image late, also there may be situations where the other specialist does not prioritize the patient despite the report of the radiologist, thus causing the delay of treatment. These situations can harm the patient's body and soul integrity. It has been discussed under separate headings who will be responsible for the crime of discrimination and the harm caused by the discrimination of artificial intelligence.

b. **Prioritization of Admission to Hospital**

The main reason why artificial intelligence discriminates is that different categories are included in the data entered into the artificial intelligence. Different categories are defined for artificial intelligence. Artificial intelligence is required to make decisions according to the specified categories. In order to avoid discrimination here, data entry should be provided without creating a category. However, some special category groups are more in need of protection than other categories. In this case, there will be a possibility that it will not be able to provide protection to individuals in need of protection before a category is created. While it is necessary to create a category in order to provide the best health service to individuals, on the other hand, the same situation can cause discrimination (Cofone 2019: 1389). One of these categories is the patient groups that should be given priority for hospitalization. Here, algorithmic decision-making mechanisms are used to decide which patient should be admitted to hospital while others are rejected. For example, there are health institutions where artificial intelligence is used to identify risky patients and to be treated by hospitalization. These institutions aimed to determine the risk of death of pneumonia patients by artificial intelligence and to provide emergency hospitalization for patients in the risk group (Orwat 2020: 40). It is expected that the algorithm will group and prioritize the patient, which it predicts will result in a better outcome for a particular disease, to be hospitalized and to start the treatment. However, there are studies showing that algorithms discriminate against black and minority ethnic patients (Morley et al. 2019: 6). Algorithms, which are expected to give priority to risky groups by determining genetic risk profiles, may discriminate in health care by ignoring risk values (Garattini et al. 2019: 76). Similarly, the discrimination of artificial intelligence in determining the patient profiles in infectious diseases such as COVID-19 and deciding on the patients who will be given priority hospitalization can cause death. If a patient who should be included in the priority group among the patients admitted to the hospital is not included in the priority group and encounters discrimination, it should be discussed who will be responsible on what legal basis. Criminal liability due to the violation of the prohibition of discrimination and who will be liable for legal responsibility for the harms caused by discrimination should be examined by considering different cases.

10.3 Liability for Decisions Made Due to Artificial Intelligence

It is necessary to determine who will be responsible if the patient is harmed due to the prioritization of the patient, which should actually have priority, in the imaging prioritization made based on false information about the patient, especially as a result of faulty imaging by artificial intelligence algorithms. There is no special regulation that can be directly applied to both the legal and criminal liability of artificial intelligence (Dülger 2018: 84; Akkurt 2019: 53). For this reason, the liability of artificial intelligence, of the doctor, of the operator, and the liability of the manufacturer or software developer of the artificial intelligence program have been evaluated within the scope of current regulations. In addition, it has been discussed who will be responsible for the death or harm to his health due to the fact that the patient, who has a higher priority in the prioritization of admission to the hospital, is not determined by the artificial intelligence as a priority.

a. **Physician's Responsibility**

Doctors have responsibilities due to their faults within the scope of malpractice. If the doctor has not received sufficient information from the patient and this situation causes harm to the body integrity of the patient, the physician is considered to be at fault. In addition, if the patient is harmed due to not being diagnosed correctly, the doctor is considered to be at fault (Yördem, 2019: 132). The patient may be harmed due to the treatment applied by the doctor based on faulty imaging using artificial intelligence. In this case, it is debatable whether the doctor can avoid responsibility by arguing that the misdiagnosis is caused by false imaging, namely artificial intelligence. The doctor should be considered at fault where it is expected that the doctor would have noticed the error in the imaging based on this information if he had received sufficient information from the patient. On the other hand, it can be thought that the doctor should not be responsible in cases where this information would not allow the doctor to notice the error in the imaging, even if he had received sufficient information from the patient. However, in this case, it should be examined whether the doctor performed the necessary physical examination. The doctor is considered wrong if he does not perform adequate physical examination or necessary examinations (Yördem, 2019: 132). The doctor, who seems to have noticed the error in the imaging if he had done enough physical examination or the necessary examinations, cannot be relieved of his responsibility. In this case, he should be responsible for the harm caused by his fault. The responsibility of the doctor here is the liability of fault.

It is discussed whether the doctor will be responsible for the harms that may occur in the patient's health due to the fact that the artificial intelligence does not give priority to the patient, which should be a priority for imaging. In cases where it is understood that artificial intelligence would not comply with the decision of not giving priority to a patient who should be prioritized, if the doctor received incomplete information or did not perform an incomplete examination or examination, the doctor should be responsible. In particular, individuals in the category who need treatment primarily because of their illness should have immediate access to health care. It is necessary for the doctor to immediately apply the medical treatment and in this way the right to life or the right to health should be protected. Prioritization in categories where both the right to life and the integrity of the soul and body are in danger stems from the fact that these rights are human rights. Being aware of the Hippocratic oath, doctors should pay attention and care in determining the priority categories both in imaging and in hospitalization. It is not possible for the doctor who does not show this attention and care to evade responsibility by putting forward artificial intelligence algorithms. As a matter of fact, artificial intelligence algorithms only give an idea about the patients who need priority hospitalization or the patients who should be prioritized for imaging (Price II 2017: 12). Ultimately, the decision-maker is not the artificial intelligence, but the doctor himself. The decision of artificial intelligence on whether to include the patient in the priority category is only a recommendation and is not binding. The doctor is not dependent on the prioritization decision made by artificial intelligence. Considering the decision made by the artificial intelligence, the doctor should decide whether it is a priority by evaluating both the information he received from the patient and the physical examination and other examinations he made. Ultimately, since it is the doctor who makes the decision, if discrimination has been made as a result of this decision, the doctor himself should be responsible for the harm that may arise. However, there may be cases where the algorithm decision is dominant in the doctor's decision. Although the doctor receives the necessary information from the patient and performs the necessary physical examinations and tests, these data are not the main factor in the decision of the doctor whether the patient is a priority, but imaging findings using artificial intelligence technology may be the main factor in the decision of the doctor. The patient may be harmed by the functions of making and managing inferences from patient-related data and general biomedical information, as well as the decision made by the decision support software provided to physicians (Brown & Miller, 2014: 711). If the decision made by this software had not been and the doctor would not have made that decision, the doctor should be able to avoid responsibility here. As a matter of fact, according to the characteristics of the concrete case, if the decision made by the algorithm had not been and the doctor would not have made that decision, that is, if the result of the artificial intelligence was the main factor in the doctor's decision despite the doctor fulfilling all his obligations, the doctor should be freed from responsibility. Because if the algorithm did not show that result, the doctor would not make that decision. Another situation that the doctor should not be held responsible for is that the information obtained from the patient or the data obtained from the examination and examination results, apart from the algorithm decision, is not at a level that requires the doctor to decide on the priority of the patient. In this case, the algorithm may not have included the patient in the priority category by making the necessary evaluation, while it should have decided that the patient should be prioritized. In this case, the doctor could not determine the existence of a condition that would require priority treatment for the patient, based on both other data and the data provided by artificial intelligence. Therefore, the doctor should not be responsible. If it is accepted that the doctor discriminates, the doctors will try to have all kinds of examinations in order not to take risks, and this will increase the cost. In addition, it will take a lot of time to carry out many tests and evaluate the results. In this case, the patient's health will be endangered due to waiting for the results of the examination, for which priority and immediate intervention is required. Expanding the responsibility of doctors in the ordinary course of life in this way will put stress on doctors and cause them to not be able to perform their profession properly. In addition, the doctor must avoid responsibility by fulfilling all his obligations and proving that he is not in a position to know that he is discriminating against the patient by deciding that although he can obtain the necessary information from the patient, perform the necessary physical treatment and perform the examinations, he cannot reach the data that will require the patient to be a priority, and that the artificial intelligence is not a priority for the patient.

b. **Prioritization of Admission to Hospital**

It can be a person who discriminates, or it can be software programmed by a person. Software can make discriminatory decisions even though the person has no intention or mean to discriminate while programming the software (Packin and Lev-Aretz 2018: 88). The presence of biases or deficiencies in the data used while programming artificial intelligence may unintentionally discriminate against certain groups or certain people (Yılmaz, 2020:40). For example, there are studies showing that racial prejudices can cause discrimination in doctor's decisions, albeit involuntarily (Oliver et al. 2014: 181) (Mende-Siedlecki et al. 2019: 863). This type of data set entry can also cause artificial intelligence to discriminate. Will is assessed under human dignity (Doğan, 2022: 18), and a concept related to personality (Kara Kılıçarslan 2019: 371). Artificial intelligence has not yet been given any personality. In 2017, European Parliament Report with recommendations to the Commission on Civil Law Rules on Robotics, Article 59/f' upon "the most sophisticated autonomous robots could be established as having the status of electronic persons responsible for making good any harm they may cause", it was stated that artificial intelligence or robots can be given electronic personality (Delvaux 2017). However, this report is not binding. There is not yet a country that recognizes an electronic personality for artificial intelligence.

Punishment of persons in criminal law is based on the fact that individuals are responsible for the consequences of their actions of their own free will (Caşın et al. 2021: 11). Except for the crimes that can be committed by negligence in the law, the presence of "intention" is required for the crime to be committed (Özbek and Özbek 2019: 613). For a crime to be committed intentionally, it must be committed voluntarily. In other words, in order to be able to talk about the act that caused the crime in criminal law, the behavior must be done voluntarily (Köken 2021: 263). Willpower can be defined as the capacity to actively decide what to do (Caşın et al. 2021: 6). Despite the fact that artificial intelligence acts with its own free thought, it is not actively making decisions, but automatically reacting to stimuli (Caşın et al. 2021: 6). Although artificial intelligence has been developed within the framework of the cognitive features of human beings, the aesthetic perception that is effective in the voluntary decisions of the human brain will probably never be realized for artificial intelligence (Arf 1959: 103). The ability of artificial intelligence to learn and make decisions like humans does not mean that it has free will (Erdoğan 2021: 164). Since artificial intelligence is considered to have no will and has no personality, it should be accepted that artificial intelligence itself is not criminally liable in case of discrimination.

Criminal liability arising from negligent behavior rather than a voluntarily performed behavior of artificial intelligence is also evaluated. Because, as a result of negligence, that is, the inaction of artificial intelligence, a crime may occur. For example, the patient may die because the artificial intelligence, which is supposed to give medicine to the patient at regular intervals, does not fulfill its obligation to give medicine. A patient who artificial intelligence does not prioritize by discriminating while it should give priority may die due to delay in treatment. In this case, the artificial intelligence will not be responsible, since the artificial intelligence is not recognized as a personality. For the crime of willful killing due to negligence, in the presence of other conditions specified in Article 83 of the Turkish Penal Code, the person who committed the negligence must be identified in order to determine the responsible person (Özbek and Özbek 2019: 612).

c. **The Responsibility of the Operator and the Developer**

In order to determine who is at fault regarding the harm caused by the discrimination of artificial intelligence, it is necessary to investigate whether there is an error in the codes of the software, whether it carries the standards required in such software, whether software updates are made or the lack of detection of the software or similar information (Erdoğan 2021: 153). Since discrimination is prohibited under Article 122 of the Turkish Penal Code, if the operator or software developer is at fault, they should be punished with imprisonment from one year to three years.

If he is legally liable for discrimination, in case the operator or software developer is at fault, he must compensate for the damage arising from this defect within the scope of Article 49 of the Turkish Code of Obligations No. 6098. When evaluated within the scope of the prohibition of discrimination, considering the possibility of discrimination by artificial intelligence, the data that will cause this should be arranged in a way to prevent discrimination in data entry, and the artificial intelligence should be operated and programmed in a way that does not discriminate. It should be accepted that if the software developer programs the artificial intelligence without paying attention to the prohibition of discrimination, it is negligence even if it is not intentional. The software developer should be deemed imperfect due to this negligence and should be responsible for the damage arising from discrimination caused by artificial intelligence in accordance with Article 49 of the Turkish Code of Obligations. Similarly, even if the operator does not intend to discriminate, he is obliged to inspect whether the artificial intelligence system he uses makes discrimination. The operator who does not fulfill this obligation should be considered imperfect due to his negligence. In accordance with Article 49 of the Turkish Code of Obligations, it must be liable for the damage caused by discrimination caused by artificial intelligence. The operator and the software developer are jointly and severally liable. There may be cases where the operator is a different person and the user of the artificial intelligence is different people. In this case, the user should be jointly and severally liable with the operator and the software developer.

On the other hand, artificial intelligence can be used as a tool to commit crimes, since artificial intelligence is considered as a kind of article regarding criminal liability (Aksoy 2021: 15). In this case, it is accepted that the operator or software developer who uses artificial intelligence as a means of committing a crime violates the prohibition of discrimination (Köken 2021: 267). In this case, if there is a fault, both the operator and the software developer may be sentenced to imprisonment from one year to three years, as they prevent the person from benefiting from a certain service offered to the public within the scope of Article 122 of the Turkish Penal Code.

In some cases, it has been revealed that algorithms produce discriminatory results in decision-making processes, even if the algorithms or individuals have no purpose to discriminate (Cofone 2019: 1396). Decision-making mechanisms using artificial intelligence carry the risk of causing social injustice by systematizing discrimination (Packin &and Lev-Aretz 2018: 88). In other words, systems in which artificial intelligence is used may carry the risk of discrimination due to their nature. In case of discrimination, the person who caused the damage must be responsible for his fault. However, in some cases, people who are not at fault may need to compensate the damage (Bak 2018: 220). Since the systems in which artificial intelligence is used carry the risk of discrimination, it should be accepted that those who use these systems take the risk of harm. There is a principle that those who benefit from their blessings bear their burdens. Pursuant to this principle, an operator seeking to make a profit by taking the risk of discrimination by the artificial intelligence system must bear the loss caused by discrimination. This case can be evaluated within the scope of the responsibility of the performance assistant regulated in Article 116 of the Turkish Code of Obligations, if an electronic personality is given to the artificial intelligence (Benli and Şenel 2020: 323). In this case, since there is a contractual relationship between the patient and the hospital, the hospital, as the operator, may be held liable for the damage caused by discrimination caused by artificial intelligence, within the scope of Article 116 of the Turkish Code of Obligations. However, in the current circumstances, the hospital should not be held responsible according to Article 116, since the artificial intelligence has not yet been given a personality (Ercan 2019: 39). The damage caused by the artificial intelligence used in the hospital should be evaluated under the responsibility of the organization regulated in the 3rd paragraph of Article 66 of the Turkish Code of Obligations. Accordingly, the operator is responsible for the damage caused by the activities of the enterprise. As the operator, the hospital should be held responsible for the damage caused by the discrimination of artificial intelligence in accordance with the responsibility of the organization.

Here, damage to the patient's health due to discrimination is included in the scope of damage to body integrity. Here, due to discrimination, the physical or mental integrity of the patient may be disrupted and both material and moral harm may occur (Şahin 2011: 126). Damage to body integrity is specifically regulated in Articles

54 and 56 of the Turkish Code of Obligations. According to the article, financial compensation is categorized as bodily harm, treatment expenses, loss of earnings, losses due to reduction or loss of working power, and losses due to loss of economic future. In addition, in accordance with Article 56 of the Turkish Code of Obligations, in case of damage to the bodily integrity, an appropriate amount of money may be required to be paid as non-pecuniary compensation. However, if harm to body is severe or the patient dies, the judge may also order that the relatives of the patient be paid non-pecuniary damages. Regardless of the type of damage, compensation for material and moral damage arising from the helper's act is in question (Gültekin 2018: 398). The hospital using artificial intelligence should be liable within the scope of organizational responsibility for the damage caused by discrimination against the patient.

d. **Manufacturer's Responsibility**

In cases such as the deterioration, disruption, or out of control of artificial intelligence, the artificial intelligence manufacturer is responsible. Although there are those who argue that artificial intelligence software should be considered as a "service" (Miller and Miller 2007: 428), computer software is described as a "product" in court decisions (Winter v. G.P. Putnam's Sons 1991). For damages caused by medical software, the patient must be compensated within the scope of product liability (Miller and Miller 2007: 426). Especially in case of discrimination by not detecting a disease that artificial intelligence software should detect in imaging devices, the patient should contact the manufacturer, not the hospital, within the scope of product liability (Miller and Miller 2007: 426). Within the scope of product liability, the hospital may also apply to the manufacturer based on contractual responsibility and should be able to request the elimination of defects in artificial intelligence software that cause discrimination. The fact that the artificial intelligence device discriminates among patients shows that the software is faulty. Here, there is a contractual relationship between the hospital and the manufacturer of the artificial intelligence software. Within the scope of this contractual relationship, the manufacturer is responsible for the defect as regulated in Article 219 of the Turkish Code of Obligations and the following. For this reason, the hospital should be able to use its optional rights arising from the defect against the manufacturer. Apart from the sales contract, the hospital should be able to request the elimination of the situation that causes discrimination in the artificial intelligence algorithm, within the scope of the warranty contract.

In article 3 of the Product Safety and Technical Regulations Law No. 7223, "safe product", "product with risk", and "product with serious risk" are defined. Accordingly, a safe product is a product that does not carry any risk or carries the minimum risk specific to the use of the product and provides the necessary level of protection for human health and safety, when used in accordance with the instructions for life, commissioning, installation, use, maintenance and supervision and under normal conditions of use. Here, this is the scenario, where the manufacturer is liable for damage caused by a product that does not meet the safety expectations of the average reasonable user for the product (Aldemir Toprak and Westphalen 2021:744,

745). In accordance with Article 6 of the Law No. 7223, the manufacturer or the importer is held responsible for the damage caused by the product. Here, the responsibility of the manufacturer or importer is tort liability in accordance with Article 49 of the Turkish Code of Obligations No. 6098. The injured party must prove that the product is faulty and therefore the damage has occurred (Aldemir Toprak and Westphalen 2021: 745, 746). Under product responsibility, the product must be defective when placed on the market. The manufacturer is held responsible for the damages caused by this defect in the person or property (Kulular Ibrahim 2021: 177). On the basis of holding, the manufacturer responsible for the damage caused by the active behavior or negligence of artificial intelligence, the manufacturer of the artificial intelligence software is held responsible for not being able to foresee or prevent the harm from occurring (Cetıngul 2021: 1028, 1029).

10.4 Conclusion

The use of artificial intelligence in the health sector causes discrimination among patients (Hoffman and Podgurski 2020: 6). Effective results in the fight against discrimination can be achieved by using artificial intelligence and providing machine learning with data that comply with legally determined criteria. By identifying those most in need with artificial intelligence, limited health resources can be used in the most efficient and fair way (Hoffman and Podgurski 2020: 49). Especially during the COVID-19 period, due to the lack of sufficient places in hospitals, it was decided which of the patients who applied to the hospital would be rejected and which patient would be given priority over the others by using artificial intelligence algorithms. Artificial intelligence algorithms make the decision about which patient will be hospitalized. When these decisions are examined, it was determined that the patient groups were not hospitalized by the artificial intelligence algorithm even though they needed to be hospitalized. The artificial intelligence algorithm thus discriminated against certain groups. Both the right to life and the right to health of the patients who are not given priority in treatment by discriminating have been violated. In this way, the most basic human rights have been violated in the decisions made by artificial intelligence algorithms. Criminal responsibilities and legal responsibilities of artificial intelligence algorithms for violating the prohibition of discrimination have been evaluated. Since artificial intelligence algorithms do not have personalities, they cannot be held personally responsible. In the European Report with recommendations to the Commission on Civil Law Rules on Robotics, it is stated that there should be a different type of strict liability for the damage caused by artificial intelligence from the current regulations (Delvaux 2017). By using artificial intelligence in the health sector, it has been examined who will be responsible for the damage caused by discrimination among patients. Among the existing strict liability types, the operator, software developer, or user should be held responsible within the scope of organizational responsibility. At the same time, as the manufacturer of artificial intelligence, the software developer should be held responsible for not providing the expected security from the product within the scope of tort liability in accordance with Law No. 7223. Here, artificial intelligence does not provide the necessary level of protection for human health and safety by making discriminatory decisions. Since a safe product is defined as "a product that provides the necessary level of protection for human health and safety" within the scope of Law No. 7223, artificial intelligence that violates the prohibition of discrimination is not a safe product because it harms human health. The physician, on the other hand, is responsible for the damage caused by discrimination, as he must make the necessary physical examination, perform the necessary examinations and make a decision by taking the necessary information from the patient within the scope of the obligation to inform the patient. It is expected that the doctor would have noticed the error in the imaging if he had done the physical examination or the necessary examinations. Failure of the doctor to fulfill his/her due care and attention means that he/she is at fault. However, there are exceptional cases where the doctor may be relieved of responsibility, depending on the characteristics of the concrete case. However, the main purpose is to prevent harm by preventing discrimination. For this reason, it is necessary to raise awareness of doctors that they should ensure equal access to the most basic human rights, the right to life and health rights. A high level of care and attention should be paid to the use of artificial intelligence algorithms in the health sector. Software developers should be more careful about the absence of discriminatory elements in the data sets entered into artificial intelligence. Health institutions should carefully carry out the necessary investment and R&D studies for the use of artificial intelligence technologies in the health sector. Only in this way, through raising awareness of everyone as a society, equal access to the right to health, which is one of the basic human rights, will be provided without discrimination.

References

Akkurt SS (2019) Legal liability arising from autonomous behavior of artificial intelligence. Uyuşmazlık Mahkemesi Dergisi 7(13):39–59

Aksoy H (2021) Artificial intelligence assets and criminal law. Int J Econ, Polit, Humlties Soc Sci 4(1):10–27

Aldemir Toprak IB, Westphalen FG (2021) Reflections on product liability law regarding the malfunctioning of artificial intelligence in light of the commission's COM (2020) 64 final report. Marmara Üniversitesi Hukuk Fakültesi Hukuk Araştırmaları Dergisi 27(1):741–753

Arf C (1959) Can a machine think? If so, how? In: Atatürk Üniversitesi 1958–1959 Öğretim Yılı Halk Konferansları, vol 1. Atatürk Üniversitesi, Erzurum, pp 91–103

Bak B (2018) The legal status of artificial intelligence with regard to civil law and the liability thereof. Türkiye Adalet Akademisi Dergisi 9(35):211–232

Benli E, Şenel G (2020) Artificial intelligence and tort law. ASBÜ Hukuk Fakültesi Dergisi 2(2):296–336

Brown SH, Miller RA (2014) Legal and regulatory issues related to the use of clinical software in health care delivery. In: Greenes RA (ed) Clinical decision support, the road to broad adoption. Academic Press, pp 711–740

Caşın MH, Al D, Başkır ND (2021) Criminal liability problem arising from artificial intelligence and robotic actions. Ankara Barosu Dergisi 79(1):1–74

Cetıngul N (2021) Deliberations on the legal status of artificial intelligence in terms of criminal liability. İstanbul Ticaret Üniversitesi 20(41):1015–1042

Cofone IN (2019) Algorithmic discrimination is an information problem. Hastings Law J 70(6):1389–1443

Criado N, Such JM (2019) Digital discrimination. In: Yeung K, Lodge M (eds) Algorithmic regulation. Oxford Scholarship Online

Delvaux M (2017) European Parliament. Retrieved December 15, 2021, from Report with Recommendations to the Commission on Civil Law Rules on Robotics (2015/2103(INL)). https://www.europarl.europa.eu/doceo/document/A-8-2017-0005_EN.html

Deo RC, Nallamothu BK (2016) Learning about machine learning: the promise and pitfalls of big data and the electronic health record. Circ: Cardiovasc Qual Outcomes 9(6):618–620

Doğan B (2022) A constitutional right under comparative law: right to data protection. Adalet, Ankara

Dülger MV (2018) Reflection of artificial intelligence entities on the law world: how the legal status of these entities should be determined? (M. Dinç, Ed.) Terazi Hukuk Dergisi 13(142):82–87

Ercan C (2019) Legal responsibility resulting from the action of robots solutions for non-contractual liability. Türkiye Adalet Akademisi Dergisi 11(40):19–51

Erdoğan G (2021) An overview of artificial intelligence and its law. Adalet Dergisi 148(66):117–192

Garattini C, Raffle J, Aisyah DN, Sartain F, Kozlakidis Z (2019) Big data analytics, infectious diseases and associated ethical impacts. Philos & Technol:69–85

Gültekin F (2018) Liability of the debtor due to the actions of the third party. Türkiye Adalet Akademisi Dergisi 9(35):375–403

Hoffman S, Podgurski A (2020) Artificial intelligence and discrimination in health care. Yale J Health Policy Law Ethics 19(3):1–49

IBM Research Editorial Staff (2017, April 5) IBM. Retrieved May 13, 2021, from Using AI and Science to Predict Heart Failure:https://www.ibm.com/blogs/research/2017/04/using-ai-to-predict-heart-failure/

Kara Kılıçarslan S (2019) Legal status of artificial intelligence and debates on its legal personality. Yıldırım Beyazıt Hukuk Dergisi 4(2):363–389

Köken E (2021) Criminal liability of artificial intelligence. Türkiye Adalet Akademisi Dergisi 12(47):247–286

Kulular Ibrahim MA (2021) The negative aspect of technological developments: planned obsolescence from legal perspective. Adalet, Ankara

Locklear M (2017) Engadget. Retrieved October 17, 2021, from IBM's Watson is really good at creating cancer treatment plans: https://www.engadget.com/2017-06-01-ibm-watson-cancer-treatment-plans.html

McCollough CH, Leng S (2020) Use of Artificial intelligence in computed tomography dose optimisation. ICRP 49:113–125

Mende-Siedlecki P, Qu-Lee J, Backer R, Van Bavel JJ (2019) Perceptual contributions to racial bias in pain recognition. J Exp Psychol 148(5):863–889

Miliard M (2018) GE launches New Edison platform with AI Apps. Healthcare IT News. Retrieved December 15, 2021, fromhttps://www.healthcareitnews.com/news/ge-launches-new-edison-platform-ai-apps

Miller RA, Miller SM (2007) Legal and regulatory issues related to the use of clinical software in health care delivery. In: Greenes RA (ed) Clinical decision support: the road ahead. Elsevier Inc., London, pp 423–444

Morley J, Machado CC, Burr C, Cowls J, Joshi I, Taddeo M, Floridi L (2019) The debate on the ethics of AI in health care: a reconstruction and critical review. SSRN, 1–35

Ng K, Steinhubl SR, deFilippi C, Dey S, Stewart WF (2016) Early detection of heart failure using electronic health records. Circulation: Cardiovasc Qual Outcomes 9(6):649–658

Oliver MN, Wells KM, Joy-Gaba JA, Hawkins CB, Nosek BA (2014) Do physicians' implicit views of African Americans. J Am Board Fam Med 27(2):177–188

Orwat C (2020) Risks of discrimination through the use of algorithms. Federal Anti-Discrimination Agency, Berlin

Özbek C, Özbek VÖ (2019) Determining criminal liability in artificial intelligence crimes. Ceza Hukuku Dergisi 14(41):603–622

Packin NG, Lev-Aretz Y (2018) Learning algorithms and discrimination. In: Barfield W, Pagallo U (eds) Research handbook of artificial intelligence and law. Edward Elgar Publishing, Cheltenham, pp 88–113

Price WN II (2017) Artificial intelligence in health care: applications and legal implications. The SciTech Lawyer 14(1):10–13

Roberts D, Luce E (2003) As service industries go global more white collar jobs follow. Retrieved October 22, 2021, from The New York Times:https://archive.nytimes.com/www.nytimes.com/financialtimes/business/FT1059479146446.html

Şahin A (2011) Damages for Breach of the Integrity of the Body. Gazi Üniversitesi Hukuk Fakültesi Dergisi 15(2):123–165

Sarı O (2020) Liability arising from damages caused by artificial intelligence. Union Turk Bar Assoc Rev (147):251–312

Suleyman M, King D (2019) DeepMind. Retrieved September 25, 2021, from using AI to give doctors a 48-Hour Head Start on Life-Threatening Illness:https://deepmind.com/blog/article/predicting-patient-deterioration

Topol EJ (2019) High-performance medicine: the convergence of human and artificial intelligence. Nat Med (25):44–56

Winter v. G.P. Putnam's Sons (1991) 89–16308 (United States Court of Appeals, Ninth Circuit July 12, 1991). Retrieved May 03, 2021, fromhttps://h2o.law.harvard.edu/cases/5449

Yılmaz G (2020) The European code of ethics on the use of artificial intelligence in jurisdictions. Marmara Avrupa Araştırmaları Dergisi 28(1):27–55

Yördem Y (2019) Legal responsibility of physician due to improper medical practice. Türkiye Adalet Akademisi Dergisi 11(39)

Merve Ayşegül Kulular İbrahim Who did a double major at TOBB University of Economics and Technology, graduated from the Faculty of Law in 2013 and from the Department of History in 2015. She completed his first master's degree at Queen Mary University in the United Kingdom in 2015 with the thesis titled "Protection of Privacy and Personal Data in the Absence of "The Code": The Case of Turkey." She completed her second master's degree at Hacettepe University in 2019 with her thesis titled "The Legal Foundations of Modern Technological Developments in the Ottoman Period: The Telegraph Example". She completed her doctorate in Social Sciences University of Ankara, Department of IT Law/Cyber Law in 2021 with her thesis titled "A Legal Overview of Planned Obsolescence". She is currently working as a lecturer at Social Sciences University of Ankara Faculty of Law. She was a Visiting Research Associate at the Faculty of Law, Murdoch University, Perth, Western Australia in 2020. She completed her Ph.D. at Queen's University, Ontario, Canada.

Chapter 11
The Impact of Artificial Intelligence on Social Rights

Cenk Konukpay

Abstract Digitalization has led to an increase in the use of artificial intelligence [AI] systems in many areas related to social and economic rights. AI is of significant benefit to the welfare society. Thanks to large-scale data analysis, it becomes easier to identify deficiencies in the implementation of social policies and the allocation of social benefits. However, data-driven tools may also create risks to the access to private and public services and the enjoyment of essential social rights. This is also bound up with the principle of equality and non-discrimination. Due to the increasing use of AI systems, the potential risks arise in various areas. Besides a serious transformation in the labor market with the AI technology, algorithms are used to measure the performance of employees and manage all stages of the employment including recruitment process. In addition to its impact on the right to work, AI systems are also deployed in the context of accessing social services. The use of AI in order to verify identities, prevent fraud, and calculate benefits may limit the enjoyment of social rights and discriminate against vulnerable groups of society. For this reason, there is a need to examine the place of social rights in the application of AI technology. This study seeks to carry out an analysis on how AI affects social and economic rights in the light of the principle of non-discrimination. To achieve this aim, relevant legislation and court decisions will also be examined.

Keywords Artificial intelligence · Non-discrimination · Social rights

11.1 Introduction

Artificial intelligence (AI) technologies are increasingly being used in various sectors relating to social and economic rights as a result of digitalization. AI provides a lot of advantages for the welfare society. It becomes easier to identify deficiencies in the

C. Konukpay (✉)
Istanbul Bar Association, Istanbul, Turkey
e-mail: cenkkonukpay@ybklegal.com

M. Kılıç and S. Bozkuş Kahyaoğlu (eds.), *Algorithmic Discrimination and Ethical Perspective of Artificial Intelligence*, Accounting, Finance, Sustainability, Governance & Fraud: Theory and Application, https://doi.org/10.1007/978-981-99-6327-0_11

implementation of social policies and distribution of social benefits thanks to large-scale data analysis. Data-driven tools, on the other hand, may jeopardize access to both private and public services, as well as the enjoyment of basic social rights. States use big data and digital technologies to automate, detect, and occasionally punish citizens in order to provide social protection and aid (Special rapporteur on extreme poverty and human rights [Special rapporteur] 2019, p. 5). This is also bound up with the principle of equality and non-discrimination.

In many international treaties and national constitutions, states are obliged to protect and fulfill economic and social rights. International Covenant on Economic, Social and Cultural Rights [ICESCR] is one of the most important treaties among them. The ICESCR is one of two international treaties that constitute the International Bill of Human Rights. It establishes the basis for safeguarding basic economic and social rights (The Office of the High Commissioner for Human Rights 2022). Furthermore, The European Social Charter protects social and economic rights as an essential component to the Council of Europe human rights system (Council of Europe 2022). These treaties, as well as many regional and national documents, guarantee social and economic rights in a non-hierarchical manner with political and civil rights.

Economic social and cultural rights mainly include a broad range of human rights concerning food, housing, education, health, social security, work, water, and cultural life (The Office of the High Commissioner for Human Rights 2022). These rights must be safeguarded by the states. It should be also noted that fulfilling these rights frequently necessitates extensive use of resources such as budget, personnel, etc. (Niklas 2019, p. 2). As a result, providers have been incorporating big data analytics and AI technology into their processes in order to improve their social security cycle. It is clear that when these systems collide with social welfare management, various risks and benefits for the implementation of economic and social rights may arise (Niklas 2019, p. 2).

Governments have relied on AI in order to make social rights more efficient and effective, thus to better serve citizens. This is mainly based on two purposes: identifying social needs and service delivery (Niklas 2019, p. 2).

Firstly, tools are used in the development of social policies. This might be gathering evidence, completing risk assessments, cost–benefit analyses, and other tasks connected to general policymaking (Niklas 2019, p. 2). Espoo, a Finnish city, could be an example of analyzing the big data to find targeting services to citizens and minimize social exclusion (Ristimäki 2022). Another example with the same purpose is concerning food security. In Vietnam, companies are deploying AI to boost agricultural productivity. A start-up succeeded an accuracy rate between 70 and 90% for identifying crop illnesses (Tan 2020). This could be an appropriate and even strategic use of AI. In the light of these examples, it might be said that data-driven technologies could provide important developmental benefits.

Secondly, tools are mostly used for service delivery purposes, such as checking eligibility, preventing fraud, offering public assistance to individuals. (Niklas 2019, p. 2). For instance, job centers in Poland employed a profiling system in order to combat unemployment. The system divides unemployed citizens into three different categories, based on their individual situation. (Niklas et al. 2015, p. 5) However, these kinds of systems are criticized due to the lack of transparency and accountability. For this reason, the impact of such tools on social rights needs to be extensively analyzed.

This essay will deal with the following aspects of the questions of how automated systems and AI affect the distribution of public services and whether the procedures are fair and transparent. In addition to these questions, it will be also discussed to what extent AI systems influence vulnerable groups.

The aim of this study is to highlight how AI is transforming the management of social policies, and their implications for human rights. However, it will not be possible to discuss all aspects of social rights. In order to point out essential debates regarding the effect of AI on social rights, particularly social security measures and employment processes will be briefly outlined. Within the scope of the study, relevant practices and case law will be dealt with the light of several examples in different countries. The first part of the analysis will examine the administration of social protection measures. The second part of this analysis will consider employment practices.

11.2 Social Protection

Artificial intelligence can be used for different purposes in terms of implementing social security policies: (1) identity verification, (2) eligibility assessment, (3) fraud prevention and detection. In this part, the topic will be discussed in the light of these purposes.

(A) Identity Verification

Identity cards and passports have traditionally been used for identification checks. Digitalization led to the change of this practice and electronic authentication methods have started to be used. States and organizations need verification process in order to provide benefits, proving eligibility, etc. It is also worth noting that a verified identity prevents duplication, fraud and increases efficiency. In other words, digital technology has the potential to create significant cost savings for governments. This facilitates accurate targeting and beneficial for disadvantaged groups (Special rapporteur 2019, p. 5). On the other hand, these methods involve a risk of collecting too much information about individuals and violating their privacy rights.

In Ireland, a system that requires welfare claimants to prove their identities by using a government-issued ID card exists. However, applicants are required to present proof about their identity along with several pieces of documents in order to get this card. During this process, face recognition system can also be used by the Department of Employment Affairs and Social Protection ("DEASP") to check applicants' photo with the database and confirm that they have not registered with a different identity (Human Rights Watch 2021, p. 6). However, this practice has been criticized for collecting more personal data than necessary to verify the identities of citizens (Irish

Council for Civil Liberties 2019). In addition, some applicants who were adopted as children faced difficulties by trying to prove their adoption status to the DEASP. Migrants who have incomplete identity documents, and several rural residents have also struggled by registering for the online system (Human Rights Watch 2021, p. 7). In the end, Data Protection Commission and DEASP reached an agreement by which an alternative service channel must be available to citizens other than online services where the ID card is used (Government of Ireland 2021). It must be also noted that according to research, real-time facial identification systems have less accurate results when it comes to darker-skinned people (Buolamwini and Gebru 2018, p. 1). Therefore, it might be said that the use of such systems for authentication purposes may lead to serious discriminatory consequences. This requires a need for a detailed impact assessment when deploying digital identification systems.

(B) **Eligibility Assessment**

In addition to identity verification, eligibility assessment and calculation of benefits is another purpose to deploy AI systems for organizations. This reduces the need for human decisions, especially in such processes. However, in these cases, significant problems may occur, notably due to calculation errors that occur in the systems. The Ontario Auditor-General documented, in 1 year, more than one thousand errors regarding eligibility assessment and calculations within the Social Assistance Management System in Canada. The total value of irregularities was about 140 million dollars (Special rapporteur 2019, p. 7). A similar problem in Australia concerning unemployment benefits, called Robodebt, was described as "a shameful chapter" and "massive failure in public administration" in the country's social security system (Turner 2021).

As another example, a French institution, the Caisse des Allocations Familiale [CAF], which is responsible for the administration of social benefits, faced errors in an automated system that has been used to calculate benefit payments. Benefit delays and inaccuracies caused by system modifications while receiving a housing allowance. This affected between 60,000 and 120,000 people (Human Rights Watch 2021, p. 9).

This problem was mainly caused by a software update regarding the test formula. The new system has changed the calculations of income history of beneficiaries (CAF 2021). It might be said that simple errors in the algorithms can lead to wide-ranging consequences for people. This situation may result in discrimination, especially for vulnerable groups benefiting from social security system. For this reason, AI systems that are used for eligibility assessment and calculation of benefits should be sufficiently transparent. Furthermore, it should be also strictly audited considering the risks it may pose.

(C) **Fraud Prevention and Detection**

Governments have long been concerned about fraud in social security programs, since it may entail the loss of enormous amount of money. Therefore, many digital systems have placed a strong focus on the ability to match data of applicants from various sources in order to detect fraud and anomalies. However, this creates unlimited opportunities for public authorities to reach extremely disturbing levels of monitoring and data collection. (Special rapporteur 2019, p. 8).

Risk calculation is inextricably linked to the fraud prevention systems, and AI technologies have a potential to attain extremely high levels of complexity in this area. Therefore, in addition to its benefits, it must be also necessary to consider the problems that might occur while using these systems. United Nations Special Rapporteur on extreme poverty and human rights noted three crucial points regarding this issue: Firstly, evaluating an individual's rights based on projections drawn from the behavior of a general demographic group raises many concerns. Secondly, the logic involved in these technologies, as well as how they score or categorize people, are frequently kept hidden. This makes harder to hold governments and organizations accountable for potential breaches and ensure transparency. Thirdly, risk scoring and need classification might worsen existing disparities and even increase discrimination practices. (Special rapporteur 2019, p. 9).

A concrete example of these critics is the System Risk Indicator [SyRI], a program that was used by the Dutch government. This algorithmic system was developed to detect various types of fraud concerning social benefits, allowances, and taxes etc. The system may process sensitive data collected by various government agencies. It also scores applicants based on their risk of committing fraud. In 2014, the government passed a legislation enabling the use of this system. In case of an identification of an application as a fraud risk, SyRI alerts the relevant institution. Following this notification, an applicant may face an inquiry into possible criminal activity (Human Rights Watch 2021, p. 11, 12).

However, this program was widely criticized by human rights institutions and even UN Special Rapporteur involved in the case with an amicus brief (Veen 2020). After this situation, the District Court of The Hague, held that due to the large amount of data processing, including sensitive personal data and risk profiling, SyRI is actually at risk of producing discriminatory consequences based on low socio-economic status or immigration background. The Court, therefore, decided that the use of the program must be stopped immediately (Rechtbank Den Haag 2020, par. 6.93).

In its decision, the Court stated that there is no problem for the government to use technology in order to detect fraud in social benefits. However, it was emphasized that the state has a special responsibility for the implementation of new technologies. According to the Court, the data processing activity within the scope of the SyRI is contrary to the European General Data Protection Regulation [GDPR]. It has been stated that this system violates the principles of transparency, purpose limitation, and data minimization in the GDPR (Rechtbank Den Haag 2020, par. 6.96).

Furthermore, the risk model, risk indicators, and data that make up this risk model are confidential. The SyRI legislation does not contain relevant information about how the decision model of SyRI works. It does not also impose an obligation to inform data subjects about the processing activities carried out (Rechtbank Den Haag 2020, par. 6.49). Applicants have not been notified that such a risk report had been created about them. Therefore, the Court held that the data processing activity under the SyRI program led to the disproportionate limitation of the right to respect for private life, which is protected by Article 8 of the European Convention on Human Rights (Rechtbank Den Haag 2020, par. 6.95).

In line with this decision, it could be noted that particular attention must paid to AI technologies used in risk-scoring processes. In this regard, it would be useful to mention relevant provisions regarding risk scoring in the draft AI regulation prepared by the European Commission. Article 5 of the Regulation bans AI systems deployed by public authorities "*for the evaluation or classification of the trustworthiness of natural persons over a certain period of time based on their social behaviour or known or predicted personal or personality characteristics*" (European Commission 2021). However, the Proposal limits the scope of the ban with the following situations:

(i) detrimental or unfavourable treatment of certain natural persons or whole groups thereof in social contexts which are unrelated to the contexts in which the data was originally generated or collected;
(ii) detrimental or unfavourable treatment of certain natural persons or whole groups thereof that is unjustified or disproportionate to their social behaviour or its gravity;

It is, therefore, unclear whether the ban in the Article 5 of the Proposal includes scoring or profiling systems used in social security cycle. If this provision is limited to only general-purpose scoring systems, it would not be possible to prevent scoring practices as in SyRI example (Human Rights Watch 2021, p. 17).

Briefly, it can be said that AI is used for different purposes within the scope of social security practices. This situation improves application processes, and it also creates human resources and budget gains. However, some identity verification processes disproportionately limit individuals' access to services. In addition, crucial errors occur in eligibility assessment. It should be also noted that fraud prevention and scoring practices may lead to discriminatory consequences for vulnerable groups in case of a lack of transparency in these systems.

11.3 Employment

Right to work has an important place among the social rights and when it comes to employment, we may notice an increasing use of AI in various fields including recruitment and unemployment assistance.

First of all, in the context of need classification, governments deploy automated techniques to assess if unemployment assistance will be given. This may also include the determination of the level of assistance. (Special rapporteur 2019, p. 9).

There is an important example concerning this purpose of the use of AI systems in Austria. In 2018, the Austrian authorities started to use an AI program to assess unemployed person's job prospects. This algorithmic system divides people into three different groups. By doing this, the system assists job centers in determining the type of assistance of an unemployed person should get (Lohninger and Erd 2019, p. 3). There is, nevertheless, a significant danger of prejudice. The output of the system is based on age, gender, and the algorithms even check whether females are with their children or not. In general, women are given a lower rating than men. Furthermore, according to studies, the error rate is around 15%, and 50,000 applicants were incorrectly categorized each year. It is also worth noting that the system has been developed to only help job center employees in their decision-making process. However, researches have revealed that employees have become increasingly reliant on algorithmic outputs over time (Lohninger and Erd 2019, p. 3).

With regard to this practice, Epicenter Works highlight that this system reflects stigmas among the society and causes discriminatory approach. Even though the system has transparent algorithms, the determination of a job seeker's individual value still remains unclear (Lohninger and Erd 2019, p. 3).

Besides unemployment assistance, AI is also frequently used in recruitment processes. Various algorithms can be deployed by organizations to screen candidates for open positions. Candidates may be subject to pre-selection by these systems (Köksal and Konukpay 2021). Moreover, this can be supported by additional analyses from social media (Raso et al. 2018 p. 44). Thus, AI systems help to narrow candidate pool and determine the candidates to be called for the interview. In addition, as an extreme example, automated systems that evaluate candidates' word preferences, voices, gestures, and mimics during interviews are also on the agenda (Raso et al. 2018 p. 44).

Amazon's recruitment tool can be given as an example to show problems arising from these systems. The company's tool that is used in recruitment processes has given priority to male candidates for software development and other technical position. Because algorithms have been trained with the data based on the last 10 years' candidates and these data mainly have a male candidate-dominated nature. Another relevant point is that since the algorithms analyze the language in the resumes of successful candidates, this leads to a low rate of success for resumes that do not have a similar language (Dastin 2018). Thus, the effect of gender discrimination is strengthened since a male-dominated language is also preferred by the system. Therefore, it should be careful when considering the training of the algorithms in these processes and pay attention to whether the data sets used are neutral or not (Köksal and Konukpay 2021).

11.4 Problems Related to the Lack of Digital Skills

While discussing the problems in social security and employment processes mentioned in the previous sections, it should also consider the digital skills of people. The lack of digital capacity leads to discriminatory results arising from digital systems.

This is a crucial issue even in wealthy countries with high technological development. As an example of a wealthy country, in the United Kingdom, 22% of the people do not have the essential skills that they need to use digital tools in daily life. Furthermore, additional 19% of the population cannot carry out basic tasks. This includes turning on a device or opening an application (Special rapporteur 2019, p. 9). This puts people who are not qualified to use technological tools in a vulnerable position.

It should be noted that the developers of these systems do not have enough awareness regarding this issue. IT professionals mainly assume that users will easily be able to access to their documents such as a credit history. Therefore, they can easily check the errors in the systems and appeal automated decisions about them when needed. However, considering the studies mentioned above, this is mainly not the case for many people (Special rapporteur 2019, 9). For this reason, it should have a participatory method in the development and use of these systems.

Thus, people with low ability to use digital devices can be prevented from being victims of such automated processes. In the end, alternative solutions can be created, and it will be ensured that these people are not discriminated.

11.5 Conclusion

In brief, the use of AI systems is increasing in many areas touching social rights. It is undisputed that this situation brings many conveniences in the implementation of social service such as monitoring social needs of people. Besides this, the acceleration of the processes contributes to the growth of the service offered. It also ensures that the budget and human resources are used more effectively.

On the other hand, AI may be harmful for social rights. In this study, discriminatory risks have been mentioned in the context of social services and employment processes. Related to social services, first of all, it is seen that AI, when used for identity verification, can process detailed data including special categories of personal data and it can produce discriminatory results, especially in terms of immigrants or other vulnerable groups. Furthermore, in eligibility assessment activities, there are many examples where different errors against the interest of applicants occurred in the system. This situation has a potential to produce serious discrimination, considering that the people who benefit from these services are actually among vulnerable groups. In addition, fraud prevention programs lead to serious discriminatory results, notably if they create individual scores or take into account the social groups of individuals as an input in their decision-making process.

In terms of employment activities, categorization of candidates based on age, gender, etc. may constitute discrimination. Considering the increasing use of artificial intelligence in recruitment processes, all necessary measures should be taken to prevent this kind of discrimination.

In conclusion, social rights exist in order to eliminate inequalities in society. In this regard, it is necessary to prevent the reproduction of discrimination while deploying AI systems in relation to social services. Therefore, AI should only serve as a fair distribution of economic resources and prevention of social exclusion.

References

Buolamwini J, Gebru T (2018) Gender shades: intersectional accuracy disparities in commercial gender classification. In: Conference on fairness, accountability, and transparency. Retrieved 19 April 2022, from http://proceedings.mlr.press/v81/buolamwini18a/buolamwini18a.pdf

CAF-Caisse d'allocations familiales (2021) Apl of July 2021: the anomaly resolved (Apl de juillet 2021: l'anomalie résolue). Retrieved 19 April 2022, from https://www.caf.fr/allocataires/actualites/2021/apl-de-juillet-2021-l-anomalie-resolue

Council of Europe (2022) The European social charter. Retrieved 18 April 2022, from https://www.coe.int/en/web/european-social-charter

Dastin J (2018) Amazon scraps secret AI recruiting tool that showed bias against women. Reuters. Retrieved 19 April 2022, from https://www.reuters.com/article/us-amazon-com-jobs-automation-insight-idUSKCN1MK08G

European Commission (2021) Proposal for a regulation of the European Parliament and of the Council laying down harmonised rules on artificial intelligence (artificial intelligence act) and amending certain union legislative acts, COM(2021) 206 final, Retrieved 19 April 2022, from https://eur-lex.europa.eu/legal-content/EN/TXT/?qid=1623335154975&uri=CELEX%3A52021PC0206

Government of Ireland - Department of Social Protection (2021) Press release: department of social protection reaches agreement with the data protection commission in relation to SAFE registration and Public service card. Retrieved 19 April 2022, from https://www.gov.ie/en/press-release/2eb90-department-of-social-protection-reaches-agreement-with-the-data-protection-commission-in-relation-to-safe-registration-and-public-service-card/

Human Rights Watch (2021) How the EU's flawed artificial intelligence regulation endangers the social safety net: questions and answers. Retrieved 18 April 2022, from https://www.hrw.org/news/2021/11/10/how-eus-flawed-artificial-intelligence-regulation-endangers-social-safety-net

Irish Council for Civil Liberties (2019) The Public Services Card. Retrieved 19 April 2022, from https://www.iccl.ie/2019/the-public-services-card-contd/

Köksal D, Konukpay C (2021) Büyük Veri Kullanımının Temel Hak ve Özgürlükler Üzerindeki Etkileri. In: Güçlütürk O, Retornaz E (eds) Gelişen Teknolojiler ve Hukuk III: Büyük Veri., 1st edn. Onikilevha, pp 15–89

Lohninger T, Erd J (2019) Submission for the report to the UN General Assembly on digital technology, social protection and human rights Epicenter Works. Retrieved 19 April 2022, from https://www.ohchr.org/sites/default/files/Documents/Issues/Poverty/DigitalTechnology/EpicenterWorks.pdf

Niklas J (2019) Conceptualizing socio-economic rights in the discussion on artificial intelligence. https://doi.org/10.2139/ssrn.3569780

Niklas J, Sztandar-Sztanderska K, Szymielewicz K (2015) Profiling the unemployed In Poland: social and political implications of algorithmic decision making. Fundacja Panoptykon. Warsaw. Retrieved 18 April 2022, from https://panoptykon.org/sites/default/files/leadimage-biblioteka/panoptykon_profiling_report_final.pdf

Raso F, Hilligos H, Krishnamurthy V, Bavitz C, Levin K (2018) Artificial ıntelligence & human rights: opportunities & risks. Berkman Klein Center for Internet & Society Research Publication, Retrieved 19 April 2022, from http://nrs.harvard.edu/urn-3:HUL.InstRepos:38021439

Rechtbank Den Haag - The Hague District Court (2020) C/09/550982. Retrieved 19 April 2022, from https://uitspraken.rechtspraak.nl/inziendocument?id=ECLI:NL:RBDHA:2020:1878

Report of the Special rapporteur on extreme poverty and human rights (2019). Seventy-fourth session. Retrieved 19 April 2022, from https://www.ohchr.org/Documents/Issues/Poverty/A_74_48037_AdvanceUneditedVersion.docx

Ristimäki M (2022) The City of Espoo: a unique experiment with AI. Retrieved 18 April 2022, from https://www.tietoevry.com/en/success-stories/2018/the-city-of-espoo-a-unique-experiment/

Tan J (2020) Can't live with it, can't live without it? AI impacts on economic, social, and cultural rights. Retrieved 18 April 2022, from https://coconet.social/2020/ai-impacts-economic-social-cultural-rights/index.html

The Office of the High Commissioner for Human Rights (2022) Economic, social and cultural rights Retrieved 17 April 2022, from https://www.ohchr.org/en/human-rights/economic-social-cultural-rights

Turner R (2021) Robodebt condemned as a 'shameful chapter' in withering assessment by federal court judge ABC News. Retrieved 19 April 2022, from https://www.abc.net.au/news/2021-06-11/robodebt-condemned-by-federal-court-judge-as-shameful-chapter/100207674

Veen C (2020) Landmark judgment from the Netherlands on digital welfare states and human rights Open Global Rights. Retrieved 19 April 2022, from https://www.openglobalrights.org/landmark-judgment-from-netherlands-on-digital-welfare-states/

Cenk Konukpay graduated from Galatasaray University Faculty of Law in 2013 and completed his Master's degrees at the College of Europe and Université Paris 1 Panthéon-Sorbonne. He also had the opportunity to carry out a traineeship at the Secretariat of the European Commission for Democracy through Law (Venice Commission)of the Council of Europe. He is pursuing his Ph.D. degree in public law at Galatasaray University. He was admitted to the Istanbul Bar Association in 2016. He also carried out administrative duties at Istanbul Bar Association Human Rights Centre as General Secretary from 2019 to 2022. He currently practices as an attorney at law.

Chapter 12
A Review: Detection of Discrimination and Hate Speech Shared on Social Media Platforms Using Artificial Intelligence Methods

Abdülkadir Bilen

Abstract People have political views, race, language, religion, gender, etc. may face discrimination based on their status. Again, these situations can emerge as hate speech against people. Hate speech and discrimination can occur in any environment today, as well as on social media platforms such as Twitter, Instagram, Facebook, YouTube, TikTok, Snapchat recently. Twitter is a place where people share their ideas and news about themselves with their followers. To detect situations such as sexist, racist, and hate speech, recently, Twitter data have been examined and these discourses have been tried to be determined with various analysis and classification methods. While detecting these, it is done with artificial intelligence methods such as Support Vector Machines, Artificial Neural Networks, Decision Trees, Long Short-Term Memory. Considering that many information such as events, meetings, news etc. spread rapidly on social media, it is extremely important to quickly determine how people can react in discrimination and hate speech with these methods and to be able to take precautions. In the study, firstly, after determining the discrimination and hate speech, it was determined which studies were carried out in the literature about what was shared on social media platforms. In these studies, it has been determined that artificial intelligence methods are used, and the methods used are successful. Automatic detection systems for discrimination and hate speech have been developed in many languages.

Keywords Artificial intelligence · Social media analysis · Discrimination and hate speech

A. Bilen (✉)
3rd Class Police Chief, Turkish National Police, Counter Terrorism Department, Ankara, Turkey
e-mail: abdulkadir.bilen82@gmail.com

M. Kılıç and S. Bozkuş Kahyaoğlu (eds.), *Algorithmic Discrimination and Ethical Perspective of Artificial Intelligence*, Accounting, Finance, Sustainability, Governance & Fraud: Theory and Application, https://doi.org/10.1007/978-981-99-6327-0_12

12.1 Introduction

Discrimination can be faced in many areas, but the distinctions vary according to the degree of legitimacy and the perpetrator's power. Discrimination can have political, economic, and social consequences. Hate speech that emerges with discrimination such as race and gender sometimes leads to anger, social action, and lynching (Altman 2011). In recent years, discrimination and hate speech have been encountered intensively on social media platforms. Social media facilitates interaction, content exchange, collaboration, and reaching audiences. It also brings together communities that participate in people's concerns and ideas, allowing social networking among a larger group of people. The news shared on platforms such as Twitter, which has become the backchannel of television, one of the pioneers of traditional media, has become an important source of information for the public. In addition to these, many comments on discrimination are disseminated through these platforms (Bozdağ 2020). Especially in the COVID-19 epidemic that has emerged in recent years, discrimination against elderly individuals has manifested itself on social media platforms. The elderly, who were most affected by the epidemic in the curfews, were subjected to serious criticism, and most of these criticisms occurred on social media platforms. Sarcastic, insulting, threatening, and hate speech towards the elderly can be analyzed through the data obtained on Twitter (Uysa and Eren 2020).

Again, sexual assault and harassment incidents shared on these platforms expose their victims to discrimination. There is the second victimization in cases where degrading comments are made towards the victims who share their stories. Since those who read, re-share, and like these posts are also shared on these platforms, various methods analyze these data. These analyses are made by collecting publicly available data with application programming interfaces for Twitter shares (Li et al. 2021). Artificial intelligence methods are used to analyze these shares and draw meaningful results. Machine learning, one of the artificial intelligence methods, trains machines to make them smart. In particular, it has benefits such as quickly analyzing data that are too large for data analysts to interpret and extracting various patterns from it. Artificial intelligence can detect discrimination and hate speech and make decisions on predictive investigation and sentencing recommendations. Although artificial intelligence methods are promising in many areas, they also bring serious risks when misunderstood, especially in justice and law. It can also have discriminatory effects if prejudiced and discriminatory people learn from their data (Zuiderveen Borgesius 2018).

After identifying the discrimination and hate speech, the study aims to understand studies in the literature about what is shared on social media platforms. The artificial intelligence methods used in detecting discrimination and the methods' performances are evaluated, and the contributions to the literature are revealed. The advantages and disadvantages of studies on discrimination and hate speech in Turkey and the world are discussed. In the first chapter, the studies carried out in the field are discussed, and their advantages are emphasized. The second part touches on social media. The third part explains artificial intelligence techniques. In the last section, the solutions obtained in the study are presented.

12.2 Literature

Artificial intelligence methods are used in many areas, from crime analysis to cyber-physical systems that detect attacks and monitor social media platforms in studies. Many methods in the literature detect social media discourses.

A dataset consisting of 95,292 tweets that include sexist comments on Twitter was created by Jha et al. Tweets such as "Like a Man, Smart for a Woman" towards women and "A man who doesn't act like a man" were selected for men. Two models were created with support vector machines and sequence-to-sequence algorithms, and the data were classified into three groups: "Hostile Sexist", "Benevolent Sexist", and "Non-Sexist". Words such as "Trash, Hot, Bad, Stupid, Terrible" were taught to the system as hostility, and "True, Powerful, Beautiful, Better, Wonderful" were taught as good faith. In general, the SVM performed better for the "Benevolent Sexist" and "Non-Sexist" class, while the sequence-to-sequence classifier did better with the "Hostile Sexist" class. The FastText classification, on the other hand, gave better results in all three cases than the other two models. It aimed to analyze and understand the prevalence of sexism prevalent in social media (Jha and Mamidi 2017).

As a dataset, Basile et al. (2019) used 19,600 tweets consisting of tweets written in Spanish and English made up of hate speech against immigrants and women. The study aimed to first identify the hateful content and determine whether the hate speech targets a single person or a group. As a result, it has been confirmed that hate speech against women and immigrants is difficult to detect and is an area that needs to be developed with further research.

Ibrohim and Budi (2019) aimed to automatically detect the target, category, and level of hate speech and profanity language spread on social media to prevent conflicts between Indonesian citizens. In total, 16,400 tweets were analyzed. The tweets are separated as hate speech against a religion or belief, race, ethnicity or cultural tradition, physical difference or disability, gender, and profanity or inflammatory speech unrelated to these four groups. In the first experimental scenario, multi-tag classification was used to identify abusive language and hate speech, including the target, categories, and tweet level. In the second scenario, multi-label classification was used to identify abusive language and hate speech in a tweet without determining the target, categories, and level of hate speech. The random forest decision tree classification performed best in both scenarios with fast computation time and accuracy.

Lara-Cabrera et al. (2019) presented five indicators to assess an individual's risk of radicalization. These are divided into personality traits such as disappointment and introversion. Others are indicators related to the perception of discrimination for Muslims, expressing negative thoughts for western societies, attitudes and beliefs showing support for jihad, and positive ideas. Three different datasets were used to analyze the performance of these metrics. The first dataset consists of 17,410 tweets written by 112 ISIS supporters worldwide since the Paris attacks in November 2015. The second consists of 76,286 tweets collected from 142 users related to the ISIS terrorist organization during the #OpISIS operation, known as the largest operation of the Anonymous hacker group in history. One of the features of this dataset relates to the number of languages used in tweets, including English, Arabic, Russian, Turkish, and French. In the third dataset, 120 users were added to the streaming application of the Twitter platform, and 172,530 tweets were obtained from 120 randomly selected users. The metrics mentioned in these tweets were calculated, and the density distributions were examined. Then, it was determined whether there were statistically significant differences. Statistical evidence emerged that if the user is or is at risk of radicalization, they are more likely to swear, use words with negative connotations, perceive discrimination, and express positive and negative opinions about jihad and western society. Additionally, contrary to what is expected according to the introversion indicator, it has been found that radicalized users tend to write longer tweets than others.

De Saint Laurent et al. (2020) examined the role of malicious rhetoric on the masses and groups in creating and maintaining anti-immigration communities on Twitter, a social media platform. 112,789 anti-immigration and pro-immigration tweets were analyzed using data science and qualitative techniques to achieve this. Focusing on high shareability and continuous productivity aimed to shed light on the expressions and functions in social media related to migration. It aimed to understand how and when the differences in the features and content of tweets can be used by users and for what purpose. After classifying tweets on three pro-immigrations (#WithRefugees; #RefugeesWelcome; #NoBorder) and three anti-immigrations (#BuildTheWall; #IllegalAliens; #SecureTheBorder) trending topics, they were subjected to text processing, analysis of word frequencies, and clustering. It has been concluded that tweets containing anti-immigration generally resonate and that the same hashtags should be used consistently to attract attention and gain popularity on social media platforms.

Pamungkas et al. (2020) aimed to determine the characteristics of misogynistic and non-misogynistic content in social media and whether there is a relationship between abusive phenomena and languages. Two tasks were set to classify the target of misogynistic examples by dividing shared content into misogynistic and non-misogynistic and misogynistic content into five different misogynistic behaviors. These categories are stereotype and objectification, dominance, provocation, sexual harassment and threat of violence, and disrepute. The target classification was evaluated in the active category when misogyny targets the individual and the passive category when it targets the group. 83 million tweets in English, 72 million tweets in Spanish, and 10,000 tweets in Italian were obtained. Various experiments were conducted to detect the interaction between misogyny and related phenomena, namely sexism, hate speech, and offensive language. Lexical features such as sexist insults and women words (words synonymous with or associated with "woman") have been experimentally proven to be among the most predictive features for detecting misogyny. The models used in the study generally performed well.

Mayda et al. (2021) separated 1000 tweets belonging to different target groups as hate speech, offensive expression, and none. The study first aimed to create a data set about hate speech written in Turkish. The characters, bigrams and trigrams, word unigrams, and tweet-specific feature sets of these tweets were determined. Machine learning algorithms such as Naive Bayes (NB), Decision Tree, Random Forest, and Sequential Minimal Optimization (SMO) were used. The SMO classifier used with the feature selection method based on information gain showed the best performance.

Baydoğan and Alataş (2021) aimed to detect hate speech quickly, effectively, and automatically with artificial intelligence-based algorithms. 40,623 tweets with and without hate speech in English were used as the dataset. 5 different artificial intelligence algorithms were selected. Hate speech was estimated more than 75% correctly in all methods used.

All research in the studies is data taken from the Twitter platform. It is also important to conduct studies on detecting and analyzing the comments written on YouTube, Instagram, and other social media platforms. In the examined studies, determinations were made in many languages such as English, Spanish, Italian, and Turkish. Two of the studies focused on statistical analysis, and the other six used various artificial intelligence methods. Generally, problems such as the low number of data and finding data come to the fore. Summary information of related studies is given in Table 12.1.

12.3 Social Media

Social media applications and platforms generally include many applications such as discussion forums, wikis, podcast networks, blogs, social networking sites (Facebook, Twitter, Instagram, etc.), and picture and video-sharing platforms. Today, it has made it possible to live in virtual environments with virtual reality characters through web3 and metaverse technologies. Again, companies and institutions use these platforms for business, politics, customer relations, marketing, advertising, public relations, and recruitment. With the power of social media, mass movements, and perception management are also frequently carried out. A huge amount of data are collected on platforms such as Facebook, Twitter, Instagram, YouTube, TikTok, and LinkedIn, where people share their private lives, and people can access this data. There is an interest in monitoring public opinion on political institutions, policies, and political positions, identifying trending political issues, and managing reputation on social networks. It follows how people in different countries or cultures react to certain global events. Based on the posts on these platforms, the emotional states of people, their reactions to events, and their lifestyles are analyzed. It is difficult to analyze these activities because of the data's size, dynamics, and complexity. Three methods, such as text analysis, social network analysis, and trend analysis, are used in this field. While performing the analysis, some algorithms such as support vector machine, Bayesian classifier, and clustering are used. Statistical models are derived to produce predictions from trending topics with the models (Stieglitz et al. 2014). In

Table 12.1 Summary of discrimination and hate speech studies

References	Model	Data set	Detection	Advantage
Jha and Mamidi (2017)	SVM, Seq2Seq, FastText	Twitter	Sexism	Analyzing users' tweets as benevolent, hostile sexism, and others, and detecting the prevalence of sexism
Basile et al. (2019)	SVM, LR, FastText	Twitter	Hate speech against immigrants and women	The results confirm that discourse in microblog texts is difficult to detect and is an area for improvement
Ibrohim and Budi (2019)	SVM, NB, RF	Twitter	Hate speech and abusive	To determine the target, category, and level of the discourse
Lara-Cabrera et al. (2019)	Statistical Analysis	Twitter	Radicalism	Radicalization was evaluated by examining metrics based on keywords, and a detailed description of the datasets was presented
De Saint Laurent et al. (2020)	Statistical Analysis	Twitter	Anti-immigrant or pro-immigrant	Anti-immigrant or pro-immigrant tweets were obtained by data collection methods. As a result of the analysis, it was concluded that the individual sacrificed her individuality by following and tweeting to gain popularity and attract attention
Pamungkas et al. (2020)	SVM, RNN-LSTM, BERT,	Twitter	Misogyny	The important issues in detecting misogyny with the proposed system and interlingual classification are to detect hostility in a multilingual environment
Mayda et al. (2021)	NB, DT, SMO, RF	Twitter	Hate speech	With different machine learning algorithms, tweets were separated as hate speech, offensive expression, and none
Baydoğan and Alataş (2021)	MNB, Lib-SVM, CVFDT, DT-Part, MLP	Twitter	Hate speech	To detect hate speech quickly, effectively, and automatically with artificial intelligence-based algorithms

Multinominal Naive Bayes (MNB), Concept Adapting Very Fast Decision Tree (CVFDT), Multi-layer Perceptron (MLP)

a survey study, the relationship between the discrimination experiences reported by black American adults and their use of social media was investigated. Black Americans who are more exposed to discrimination use more social media than those less exposed. This is because they see it as platforms where they can freely share their statements to voice their complaints about the situations they are exposed to, find support, and cope with racism and discrimination (Miller et al. 2021).

12.4 Using Artificial Intelligence Methods

Artificial intelligence methods also successfully solve this issue to combat discrimination issues, as they perform their tasks faster than humans. The need to automate detecting online discrimination and hate speech has emerged. Although the perception of these texts is based on natural language processing approaches, machine learning has also been used in recent years. The natural language processing approach has disadvantages such as being complex and language dependent. There are many unsupervised learning models in the literature to detect hate speech and polarization in tweets. The artificial intelligence methods used are explained below (Pitsilis et al. 2018).

12.4.1 Natural Language Processing (NLP)

Deep learning algorithms are generally used in natural language processing problems, and successful results are obtained. Text classification, one of the natural language processing problems, is classifying sentences or words on different data sets. It is the text parsing done to determine the grammatical structure of the text. It is a mood analysis that tries to understand what the text conveys and an information extraction that helps extract concepts such as date, time, amount, event, name, etc. It finds solutions to problems such as revealing the temporal relationship in the text, creating an event, determining what type of word (noun, adjective, etc.) it is, sorting the text, translating it into any language, and automatic question answering (Küçük and Arıcı 2018).

12.4.2 Convolutional Neural Network (CNN)

An image consisting of an input, an output, and a convolutional neural network is used for operations such as classifying and clustering similarities. On the other hand, text analysis separates each character in the sentences individually and removes non-standard characters. Again, the machine is trained by dividing the sentences into words first. With the trained system, cases such as profanity and discrimination are detected (Park and Fung 2017).

12.4.3 Support Vector Machine (SVM)

A discriminant classifier is a machine learning derived from statistical learning theory based on the inherent risk minimization principle. It aims to reduce test error and computational complexity and defines a hyperplane with optimality in training data between two classes. The hyperplane achieves this optimum separation with the furthest distance from the closest training data. Binary classification is one of the supervised learning methods used to detect regression and outliers (Ansari et al. 2020).

12.4.4 Decision Tree (DT)

A decision tree is a support tool that uses a tree-like graph or decision model. Their possible outcomes include resource costs, utility, and incidental event outcomes. It displays an algorithm that contains only conditional control statements and is the recursive binary division of a dataset into subsets. This results in a tree with decision and leaf nodes. It has two or more branches that represent decisions or classifications. The top node is the root node. Then the smallest tree that fits the data is found. It is chosen because of its high stability, ease of interpretation, and ability to strengthen predictive models with accuracy (Ansari et al. 2020).

12.4.5 Logistic Regression (LR)

Text classification is a common classification method for many applications such as computer vision. It calculates the probability of a sample belonging to a certain class by learning the parameter set (Ansari et al. 2020).

12.4.6 Random Forest Classifier (RF)

A random forest is an identically distributed random vector consisting of a collection of tree-structured classifiers. One vote unit is given for each tree. It is based on the bagging algorithm and uses the ensemble learning technique. Many random trees are generated in the data subset and combine the output of all trees. Its advantage is that it gives more accurate predictions by reducing overfitting and variance compared to other decision tree models (Ansari et al. 2020).

12.4.7 Long Short-Term Memory (LSTM)

Long short-term memory units are repetitive neural networks that learn about long-term addictions. It is a repetitive neural network consisting of long short-term memory units. Its common unit consists of a cell, an input, an output, and a forgetting gate. The cell remembers values and three gates regulate the flow of information into and out of the cell. It is a deep learning model mostly used to analyze sequential data that makes time series estimation and is applied for operations such as speech recognition, music composition, and language translation (Ansari et al. 2020).

12.4.8 FastText

The FastText classification offered by Facebook AI research has proven efficient for text classification. In terms of accuracy, it is generally similar to deep learning classifiers and is faster in the training phase. It uses the word bag as features for classification and the n-gram bag, which captures partial information about local word order. According to the task, the model allows updating word vectors by backpropagation during training to fine-tune word representations (Jha and Mamidi 2017).

12.4.9 Sequence to Sequence (Seq2Seq)

The basic sequence-to-sequence model consists of an encoder and a decoder. The encoder processes the elements in the input array and compiles the captured information with a vector called context. In contrast, the encoder reads the input sequence and generates a fixed-length intermediate hidden representation. The entire sequence entering the system goes through the processing stage, and the content is sent to the decoder. After these operations are performed, the output array elements are produced. A deep learning model is successful in machine translation, image caption, and text summarization (Jha and Mamidi 2017).

12.5 Conclusion

Various research is carried out to detect discrimination such as hate speech, sexism, and anti-immigration and find solutions to these issues. These discourses are mostly on social media platforms, so research is mostly done on these platforms. In addition to research on the diversity of people and communities using social media, methods that try to understand the feelings of those who write on these platforms have come to the fore in recent years. Particularly, after tweets with opposing or biased expressions on issues such as immigration, people support this idea or give reactions showing that they are against it. Here, it aims to gain popularity or gain more followers on their profile. As a result, the desire to be included in successive groups leads one away from individuality. Artificial intelligence algorithms and statistical analysis methods such as SVM, LR, RNN, NB, RF, DT, and FastText are mostly used for detection. It has been determined that artificial intelligence methods are generally successful, but deep learning-based methods are more successful than machine learning. It is understood that artificial intelligence-based automatic identification systems detect quickly and effectively. When artificial intelligence-based decision-making systems are of interest to people and make decisions about these people, great care must be taken, and final human decision-making practices must be included. It should be considered as an auxiliary element; otherwise, it can cause irreversible results and make people suffer.

Although text analyses in English, Spanish, and Italian languages are frequently encountered in the literature, some studies in Turkish are also found. Quickly predicting the discourses in social media and what results they can achieve will increase the speed of measures against theses problems and support the institutions and organizations involved in the subject. For future studies, many more studies need to be done on topics such as hate speech in Turkish, discourses about immigrants, sexism, and discrimination. Some decision support systems can be designed to detect, prevent, and take precautions against discrimination and hate speech. In addition, it was concluded that detection and analysis should be increased for comments and correspondence on other social media platforms.

References

Altman A (2011) Discrimination. In: The Stanford encyclopedia of philosophy (Winter 2020 Edition). Edward N. Zalta

Ansari MZ, Aziz MB, Siddiqui MO, Mehra H, Singh KP (2020) Analysis of political sentiment orientations on twitter. Procedia Comput Sci 167:1821–1828

Basile V, Bosco C, Fersini E, Debora N, Patti V, Rangel F, Rosso P, Sanguinetti M (2019) Semeval-2019 task 5: Multilingual detection of hate speech against immigrants and women in twitter. In: 13th international workshop on semantic evaluation. pp. 54–63

Baydoğan VC, Alataş B (2021) Performance assessment of artificial intelligence-based algorithms for hate speech detection in online social networks. Fırat Univ J Eng Sci 33(2):745–754

Bozdağ Ç (2020) Bottom-up nationalism and discrimination on social media: an analysis of the citizenship debate about refugees in Turkey. Eur J Cult Stud 23(5):712–730

De Saint Laurent C, Glaveanu V, Chaudet C (2020) Malevolent creativity and social media: creating anti-immigration communities on Twitter. Creat Res J 32(1):66–80

Ibrohim MO, Budi I (2019) Multi-label hate speech and abusive language detection in Indonesian twitter. In: Proceedings of the third workshop on abusive language online. pp. 46–57

Jha A, Mamidi R (2017) When does a compliment become sexist? Analysis and classification of ambivalent sexism using twitter data. In: Proceedings of the second workshop on NLP and computational social science. pp. 7–16

Küçük D, Arıcı N (2018) A literature study on deep learning applications in Natural language processing. Int J Manag Inf Syst Comput Sci 2(2):76–86

Lara-Cabrera R, Gonzalez-Pardo A, Camacho D (2019) Statistical analysis of risk assessment factors and metrics to evaluate radicalisation in Twitter. Futur Gener Comput Syst 93:971–978

Li M, Turki N, Izaguirre CR, DeMahy C, Thibodeaux BL, Gage T (2021) Twitter as a tool for social movement: an analysis of feminist activism on social media communities. J Community Psychol 49(3):854–868

Mayda İ, Diri B, Yıldız T (2021) Hate speech detection with machine learning on Turkish tweets. Eur J Sci Technol 24:328–334

Miller GH, Marquez-Velarde G, Williams AA, Keith VM (2021) Discrimination and Black social media use: sites of oppression and expression. Sociol Race Ethn 7(2):247–263

Pamungkas EW, Basile V, Patti V (2020) Misogyny detection in twitter: a multilingual and cross-domain study. Inf Process Manage 57(6):102360

Park JH, Fung P (2017) One-step and two-step classification for abusive language detection on twitter. arXiv preprintarXiv:1706.01206

Pitsilis GK, Ramampiaro H, Langseth H (2018) Effective hate-speech detection in Twitter data using recurrent neural networks. Appl Intell 48(12):4730–4742

Stieglitz S, Dang-Xuan L, Bruns A, Neuberger C (2014) Social media analytics. Bus Inf Syst Eng 6(2):89–96

Uysa MT, Eren GT (2020) Discrimination against the elderly on social media during the COVID-19 epidemic: Twitter case. Electron Turk Stud 15(4)

Zuiderveen Borgesius F (2018). Discrimination, artificial intelligence, and algorithmic decision-making. Strasbg: Counc Eur, Dir Gen Democr, 49

Abdülkadir Bilen He graduated from Ankara University Computer and Instructional Technologies Department and from Police Academy in 2004. Between 2004 and 2021, he worked in different ranks in the administrative units, public order, information technologies, combating cybercrime, intelligence and criminal units of the Turkish National Police. He currently works as a Branch Manager in the Turkish National Police Counter-Terrorism Department. After his master's degree in computer engineering, he completed his doctorate education from the same department. He has studies on information security, cyber security, cybercrime analysis, criminalistics, artificial intelligence and risk analysis.

Chapter 13
The New Era: Transforming Healthcare Quality with Artificial Intelligence

Didem İncegil, İbrahim Halil Kayral, and Figen Çizmeci Şenel

Abstract Turing was one of the first and prominent names of artificial intelligence as an independent science discipline. As a result of the lectures, he gave at the London Mathematical Society, he wrote an important article called "Computing Machinery and Intelligence" in 1950, based on the idea that whether machines or robots can think, Artificial intelligence (AI) is a revolution in the healthcare industry. The primary purpose of AI applications in healthcare is to analyze the links between prevention or treatment approaches and patient outcomes. AI applications can be as simple as using natural language processing to convert clinical notes into electronic data points or as complex as a deep learning neural network performing image analysis for diagnostic support. Artificial intelligence (AI) and robotics are shown to be used in many areas in the health sector Keeping Well, Early Detection, Diagnosis, Decision Making, Treatment, End of Life Care, Research and Training. Electronic patient records and recording of observation results in electronic environment have been started with the digital transformation. One of the most important issues here is that hospitals keep patient records in their own electronic environment. In the near future, using the sensors of these devices; it is aimed to store the data collected from patients in the cloud computing environment and to use them in analysis. Despite the potential of AI in healthcare to improve diagnosis or reduce human error, a failure in an AI program will affect a large number of patients.

Keywords AI in healthcare · Mechanical man · Intelligent machines · Cybernetics · Health care · Artificial intelligence in health care · Robotics · Medical things (IoMT) · Virtual artificial intelligence · Physical artificial

D. İncegil (✉)
Business Administration at the Faculty of Economics and Administrative Sciences, Hacı Bayram Veli University, Ankara, Turkey
e-mail: didemincegil@gmail.com

İ. H. Kayral
Health Management Department, Izmir Bakircay University, Izmir, Turkey

F. Ç. Şenel
Faculty of Dentistry of Karadeniz Technical University, Trabzon, Turkey

M. Kılıç and S. Bozkuş Kahyaoğlu (eds.), *Algorithmic Discrimination and Ethical Perspective of Artificial Intelligence*, Accounting, Finance, Sustainability, Governance & Fraud: Theory and Application, https://doi.org/10.1007/978-981-99-6327-0_13

intelligence · Electronic medical record adoption model · Personal health system · Family physician information system

13.1 Introduction

Overview of Current Artificial Intelligence

The idea of intelligence and the possibility of intelligent machines has been discussed extensively by philosophers, writers, and engineers since the beginning of human history. In the thirteenth century, Ismail al-Jazzari, who is known as the founder of cybernetics, made various programmable mechanical robots such as elephant clock, water-raising machines, musical robot group, etc. (Keller 2020). Likewise, Leonardo da Vinci's mechanical knight could sit, stand, and move his arms. Descartes, while envisioning animals as automatons, was interested in "mechanical man" as a metaphor (Buchanan 2005).

Although the term "intelligent machine" is of interest to philosophers as old as Descartes, the most popular term was first proposed by Alan Turing (1950). Alan Turing was one of the first and prominent names of artificial intelligence as an independent science discipline. Alan Turing, who worked for the British Government during World War II, developed a machine called The Bombe, known as the first electro-mechanical computer working to decipher the German army's Enigma crypto (Haenlein and Kaplan 2019).

As a result of the lectures he gave at the London Mathematical Society, he wrote an important article called "Computing Machinery and Intelligence" in 1950, based on the idea that whether machines or robots can think, learn, and act like humans. In this article, Turing created a test of behavioral intelligence that is still in use to classify a system/computer as intelligent. According to the Turing Test, the computer will have some key capabilities to pass the test: natural language processing, knowledge representation, automated reasoning and machine learning. If the human interrogator cannot tell whether the answers came from a person or a computer, the computer passes the test (Turing 2012). If the term AI (artificial intelligence) has a birthdate, it is August 31, 1955. MIT researchers; Marvin Minsky and John McCarthy held a workshop at Dartmouth College and the term "artificial intelligence" was first introduced at this summer workshop (McCarthy et al. 1955). The commercialization of AI began in the 1980s and has grown into a multi-billion-dollar industry advising military and commercial interests.

In the late 1980s, AI regained public attention, especially in the United States and Japan. Many companies began to produce software, robots, expert systems, and hardware, so AI transformed into an industry in these years as the part of DARPA Strategic Computing Initiative. In 1986, Carnegie Mellon Researchers produced the first autonomous car. In 1995, IBM developed a chess-playing computer called Deep Blue, as an important step in the history of AI. Deep Blue, which has numerous mastery games in its memory, could evaluate 200 million different positions per

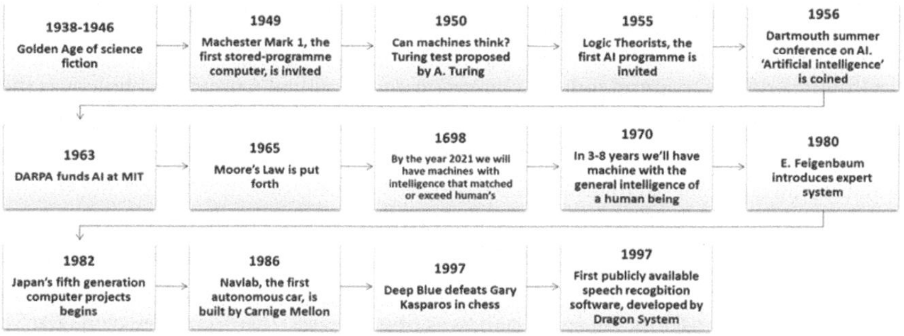

Fig. 13.1 Timeline of early AI developments (1950s to 2000). *Source* (OECD (2019), Artificial Intelligence in Society, OECD Publishing, Paris. https://doi.org/10.1787/eedfee77-en)

second (Russell and Norvig 2010). Deep Blue defeated world chess master Garry Kasparov on February 10, 1996, and Deep Blue became the first computer system to beat a human player (Fig. 13.1).

Modern AI has developed from an interest in machines that think to ones that sense, think, and act. The popular idea of AI is of a computer, hyper capable in entire domains, such as was seen even decades ago in science fiction with HAL 9000 in 2001: A Space Odyssey or aboard the USS Enterprise in the Star Trek franchise.

Lately, Rosenblatt's ideas laid the groundwork for artificial neural networks, which have come to command the field of machine learning in solving computational problems in training expressive neural networks and the ubiquity of data necessary for robust training. The resulting systems are called deep learning systems and showed important performance developments over previous generations of algorithms for some use cases (Metz 2019).

Besides healthcare, there are many industries that are more advanced in adapting AI to their workflows. AI often increases capacity, capability, and performance rather than replacing humans (Topol 2019). The self-driving car, for example, demonstrates how an AI and human might work together to succeed a goal, which improve the human experience (Hutson 2017). Another example such as legal document review, the AI working with the human reviews more documents at an advanced level of precision (Xu and Wang 2019). This team concept of human and AI is known in the AI literature as a "centaur" (Case 2018) and in the anthropology literature as a "cyborg" (Haraway 2000).

AI in Non-health Sector

Artificial intelligence although often associated with physical devices and activities, it is actually very suitable for professional activities that rely on reasoning and language. Accounting and auditing are beginning to utilize AI for repetitive task automation like accounts receivable coding and anomaly detection in audits. Engineers and architects have long applied technology to improve their design, and AI is set to speed up that trend (Noor 2017). Finance has also been an early adopter of machine learning and AI techniques. The field of quantitative analytics was born in reply to the computerization of the major trading exchanges.

Oftenly, content recommendation based on an individual's previous choices is the most visible application of AI in the media. Major distribution channels like Netflix and Amazon leverage machine learning algorithms for content recommendation to increase sales and engagement (Yu 2019).

In the music industry, startups such as Hitwizard and Hyperlive generalize these two ideas to try to predict which songs will be popular. An emerging AI quality is generative art. Software called Deep Dream, which can create art in the style of famous artists such as Vincent van Gogh, was first released (Mordvintsev et al. 2015).

Another more disturbing use of AI surfaces in the trend known as "deepfakes", technology that enables face and voice swapping in both audio and video recordings. The deepfake technique can be used to create videos of people saying and doing things that they never did, by changing their faces, bodies, and other features onto videos of people who did say or do what is shown in the video.

For the security industry, AI technology is well suited because the domain exists to detect the rare exception, and vigilance in this regard is a key strength of all computerized algorithms. In security, one current application of artificial intelligence is automatic license plate reading based on basic computer vision. Predictive policing has captured the public imagination, potentially due to popular representations in science fiction films such as Minority Report.

In the commercial sector, AI technology is increasing. AI technologies can read e-mails, chat logs, and AI-transcribed phone calls in order to identify insider trading, theft, or other abuses.

Space exploration is another area where artificial intelligence is used—an unusual and interesting one. It could be provocatively argued that our solar system is (probably) only inhabited by robots, and that one of these robots has artificial intelligence. NASA has sent robot rovers to explore the surface of Mars. NASA aboard the latest rover robot, Curiosity; It included a navigation and target acquisition AI called AEGIS (Incremental Science System Aggregation Autonomous Discovery) (Francis et al. 2017).

13.2 Artificial Intelligence in Health Care

To evaluate where and how artificial intelligence (AI) can provide improvement opportunities, it is important to understand the current context of healthcare and the drivers for change.

AI is probably to promote automation and provide context-relevant knowledge, synthesis, and recommendations to patients, and the clinical team. AI tools prioritize human labor focused on more complex tasks; can be used to reduce costs and gain efficiency; identify workflow optimization strategies; and to automate highly repetitive business and workflow processes. Also, AI tools can be used to reduce medical waste.

- failure of care delivery,
- failure of care coordination,
- overtreatment or low-value care,
- pricing failure,
- fraud and abuse,
- and administrative complexity (Becker's Healthcare 2018).

Artificial intelligence (AI) and robotics are shown to be used in many areas in the health sector.

Keeping Well

One of the most important potential benefits of artificial intelligence is helping people stay healthy so that they don't need or at least don't need a doctor as often. Using of AI and the Internet of Medical Things (IoMT) in consumer health applications is already helping people.

AI increases the ability for healthcare professionals to better understand daily patterns and needs of the people they care for, and with that understanding they can provide better feedback, guidance and support for staying healthy.

Early Detection

AI is already being used to diagnose diseases like cancer more accurately and earlier.

In the opinion of the American Cancer Society, a high percentage of mammograms give false results, leading to one in two healthy women being told they have cancer. Using of AI enables mammograms to be reviewed and translated with 99% accuracy and 30 times faster, reducing the need for unnecessary biopsies (Griffiths 2016).

Diagnosis

Watson for Health which is an IBM's product helps healthcare organizations implement cognitive technology to unlock massive amounts of health data and power diagnostics. Watson can review and store far more medical knowledge—every medical journal, symptom, and case study of treatment and intervention worldwide—exponentially faster than any human.

Google's DeepMind Health works with clinicians, researchers, and patients to solve today's health problems. Google's DeepMind Health works with clinicians, researchers, and patients to solve today's health problems.

Decision-Making

Improving healthcare requires aligning big health data with appropriate and timely decisions, and predictive analytics can support clinical decision-making and actions, as well as prioritize administrative tasks.

Using pattern identification to identify patients at risk of advancing a condition—or seeing it deteriorate due to lifestyle, environmental, genomic, or other factors—is another area where AI is beginning to take hold in healthcare.

Treatment

AI can help clinicians take a more comprehensive approach to disease management, better coordinate care plans, and better manage and adhere to patients' long-term treatment programs. For more than 30 years, robots have been used in medicine.

End-of-Life Care

Robots have the potential to revolutionize end-of-life care by helping people stay independent for longer, reducing the need for hospitalizations and nursing homes. Along with advances in humanoid design, AI is enabling robots to go even further and have "conversations" and other social interactions with people to keep aging minds sharp.

Research

The road from research lab to patient is a long and costly. According to the California Biomedical Research Association, it takes an average of 12 years for a drug to reach a patient from a research lab. Drug research and discovery is one of the newest applications for AI in healthcare.

Training

AI allows those in education to go through natural simulations in a way that simple computer-driven algorithms cannot. The advent of natural speech and the ability of an AI computer to instantly draw from a large database of scenarios mean that the response to a trainee's questions, decisions, or advice can be challenging in a way that a human cannot.

When applying these tools, it is essential to be thoughtful, fair and inclusive to avoid adverse events and unintended consequences. This requires ensuring that AI tools are compatible with users' preferences and the ultimate goals of these technologies (Baras and Baker 2009).

Driven by incentives for more population health management approaches to move to reimbursement and increase personalization, innovation in AI technologies is likely to improve patient health outcomes. Through applications, workflows, interventions, and support for distributed healthcare delivery outside of a traditional brick-and-mortar, encounter-based paradigm. The challenges of data accuracy and privacy protection will depend on whether AI technologies are classified as a medical device or as an entertainment application. These consumer-facing tools are likely to support fundamental changes in interactions between healthcare professionals and patients and their caregivers. Tools such as single-ended ECG monitoring or continuous blood glucose monitors will change the way health data are created and used (Lee and Korba 2017).

AI System Reliance on Data

Data are critical to providing evidence-based healthcare and developing any AI algorithm. Without data, the underlying characteristics of the process and results are unknown. This has been a gap in healthcare for many years, but over the past decade, key trends in this space (such as wearables) have transformed healthcare into a heterogeneous, data-rich environment (Schulte and Fry 2019).

Generating large amounts of data about an individual from a variety of sources, such as claims data, genetic information, device detection, surveillance, radiology images, intensive care unit surveillance, electronic health record care documents and medical medical information, is now common in healthcare and healthcare.

In app stores, there are more than 300,000 health apps, with more than 200 added daily, and these apps have doubled overall since 2015 (Aitken et al. 2017).

The accumulation of medical and consumer data has led patients, caregivers, and healthcare professionals to be responsible for collecting, synthesizing, and interpreting data far beyond human cognitive and decision-making capacities. AI algorithms require large volumes of training data to achieve adequate performance levels for "success" and multiple frameworks and standards exist to encourage data collection for AI use. These include standardized data representations that both lead data at rest and data in motion.

In addition, there are many instances where standardization, interoperability, and scale of data collection and transfers cannot be achieved in practice. Due to various barriers, healthcare professionals and patients cannot electronically request patient records from an outside facility after care is provided (Lye et al. 2018).

A major challenge for data integration is the lack of strict laws and regulations for the secondary use of routinely collected patient healthcare data. Most laws and regulations regarding data ownership and sharing are country-specific and based on evolving cultural expectations and norms. In 2018, a number of countries supported personal information protection guidance by moving from laws to specifications. The European Union has a strict regulatory infrastructure that prioritizes personal privacy, detailed in the General Data Protection Regulation (GDPR), which came into force on May 25, 2019 (European Commission 2018).

Differences in laws and regulations are partly the result of differing and evolving perceptions of appropriate approaches or frameworks for the ownership, administration and control of health data. There is also a lack of agreement on who can benefit from data-sharing activities. Patients may not realize that their data can be monetized through AI tools for the financial benefit of various organizations, including the organization that collects the data and the AI developers. If these issues are not adequately addressed, we risk an ethical conundrum where patient-provided data assets are used for monetary gain without express consent or compensation. Cloud computing is another particularly challenging topic, placing physical computing resources in widespread locations, sometimes across international borders. Cloud computing can cause catastrophic cybersecurity breaches as data managers try to maintain compliance with many local and national laws, regulations and legal frameworks (Kommerskollegium 2012). Finally, to truly revolutionize AI, it is critical to consider the power of connecting clinical and claims data with data beyond the narrow, traditional care setting, by capturing social determinants of health as well as other patient-generated data.

In addition to the problems with data collection, choosing an appropriate AI training data source is critical because training data influences output observations, interpretations, and recommendations. Training data; are systematically biased, for example, due to underrepresentation of individuals of a particular gender, race, age, or sexual orientation, those biases are modeled, propagated, and scaled in the resulting algorithm. The same applies to human biases (intentional and unintentional) operating in the environment, workflow, and results from which data are collected. Bias may also be present in genetic data where the majority of the sequenced DNA is obtained from people of European descent (Stanford Engineering 2019). Training AI with these biases from data sources risks generalizing it incorrectly to populations that do not represent it. In order to promote the development of effective and fair AI tools, AI developers should be based on gender, culture, race, age, ability, ethnicity, sexual orientation, socioeconomic status, privilege, etc. It is important to ensure diversity (West et al. 2019).

To begin with, system leaders should identify key areas of need where AI tools can help, help address existing disparities, and where implementation will result in better outcomes for all patients. These areas should also have an organizational structure that addresses other ethical issues such as patient–provider relationships, patient privacy, transparency, notification and consent, as well as other ethical issues such as technical development, validation, implementation, and maintenance of artificial intelligence tools. the ever-evolving learning healthcare system. Implementing healthcare AI tools requires collaboration and prioritization of governance structures and processes by information technology professionals, data scientists, ethicists and lawyers, clinicians, patients, and clinical teams and organizations (Fig. 13.2).

Under the main heading of health services; all fields related to modern health, such as medical health services, digital health, mobile health, intersect with artificial

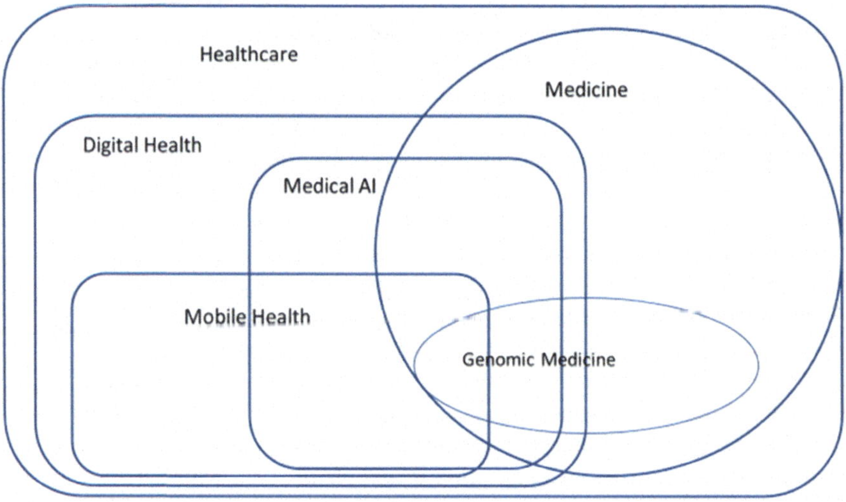

Fig. 13.2 Relationship between digital health and other terms. *Source* (Shin (2019) Modified from Choi YS. 2019 with permission. AI, artificial intelligence)

intelligence. Artificial intelligence-based solutions appear with solutions such as voice response systems in mobile health, and applications such as diagnosis and diagnosis in the field of medicine artificial intelligence applications form the basis of almost all services provided in digital health (Shin 2019).

13.3 The Future of Digitalization and Artificial Intelligence in Healthcare

Electronic patient records and recording of observation results in electronic environment have been started with the digital transformation. One of the most important issues here is that hospitals keep patient records in their own electronic environment. In the following processes, in order to be able to easily process patient data and run methods such as big data analysis and machine learning on it, by countries and even on a global basis; it is aimed to keep all electronic patient records in cloud computing infrastructures and to switch hospitals from local information infrastructures to cloud solutions (Atasoy et al. 2019).

Although there are few approved and implemented studies on the use of artificial intelligence in digital health, academic research on this subject shows that the areas of use will be much more in the near future. Machine learning and artificial intelligence-based studies, especially in the field of diagnosis; it focuses on chronic diseases such as cancer, heart diseases, Alzheimer's and kidney diseases, and it shows that in the near future, we will encounter artificial intelligence-based applications more frequently in the diagnosis processes of these diseases (Tran et al. 2019).

In chronic diseases, internet-based tracking solutions of objects are gaining popularity. In the near future, using the sensors of these devices; it is aimed to store the data collected from patients in the cloud computing environment and to use them in analysis. Especially for heart patients, by monitoring indicators such as heart rate and blood pressure, it will enable to detect cases such as heart attacks earlier (Johnson et al. 2018).

The training of artificial intelligence and its ability to produce healthy results are directly proportional to the amount and quality of the data used. Due to the use of wearable technologies and the widespread use of IoMT devices, situations such as the ability of individuals to manage their own health conditions and personalized health applications have become widespread. This type of wearable devices can also send data to the doctor of individuals, and thus, the health status of individuals can be followed. In the following process, as the data obtained by monitoring the devices increase, artificial intelligence-supported systems will become widespread.

Another important area in the virtualization processes in health is TeleHealth applications. In a period when there should be no physical communication with hospitals; virtual nursing, remote treatment, NLP-based diagnosis and referral systems are used and developed. As such applications become feasible with artificial intelligence, facts such as human relations, empathy, and the importance of communication between healthcare professionals and patients will come to the fore and quality in healthcare will gain a new dimension in the future.

13.4 The Quality in Health Care with Artificial Intelligence

Artificial intelligence (AI) and the role of machine learning in the health care sector are quickly expanding. This enlarging could be accelerated by the global spread of COVID-19, which provides new opportunities for AI prediction, screening, and image processing capabilities (McCall 2020).

AI applications can be as simple as using natural language processing to convert clinical notes into electronic data points or as complex as a deep learning neural network performing image analysis for diagnostic support. The purpose of these tools is not to replace health care professionals, but to better inform clinical decision-making to provide a better patient experience and increase clinicians' safety and reliability.

Clinicians and health systems using these new tools should be aware of some of the key issues with safety and quality in the development and use of machine learning and artificial intelligence. Inaccurate algorithm output or incorrect interpretation by clinicians can lead to significant adverse events for patients. AI used in health care, especially clinical decision support or diagnostic tools, can have a significant impact on patient treatment (Char et al. 2018).

Primary limitation in developing effective AI systems is the caliber of data. Machine learning, the underlying technology of many AI tools; It feeds data features such as patient demographics or disease status from large datasets into an algorithm to draw more accurate relationships between input characteristics and outcomes. The limitation for any AI is that the program cannot exceed the performance level reflected in the training data (Jiang 2017).

With healthcare systems starting to develop machine learning algorithms in-house and more AI applications coming to market, it's vital that providers take a thoughtful approach to applying AI to the care process.

Healthcare providers must improve their understanding of how machine learning algorithms are developed. Collecting large amounts of data has driven growth in new AI tools, but the success of these tools relies on ensuring that the data is of high quality: accurate, clinically relevant, and tested in multiple environments.

Whether AI tools are easily adopted in healthcare settings depends on building public confidence in AI. Clinicians and patients must believe that AI applications are safe and effective with a basic process that can be explained to users. Achieving this goal requires transparency from developers and clinical credibility for users (Jamieson and Goldfarb 2019).

Despite the potential of AI in healthcare to improve diagnosis or reduce human error, a failure in an AI program will affect a large number of patients. Clinical acceptance will ultimately depend on establishing a comprehensive evidence base to demonstrate safety and security.

13.5 AI in the Health Care Safety: Opportunities and Challenges

Artificial intelligence (AI) is a revolution in the healthcare industry. The primary purpose of AI applications in healthcare is to analyze the links between prevention or treatment approaches and patient outcomes. AI applications can save cost and time for diagnosis and management of disease states, making healthcare more effective and efficient. AI enables fast and comprehensive analysis of large datasets to effectively enable decision-making with speed and accuracy.

AI is often described as two types: virtual and physical. Virtual AI incorporates computing from deep learning applications such as electronic health records (EHRs) and image processing to assist physicians in diagnosing and managing disease states. Physical AI includes mechanical advances in surgery such as robotics and physical rehabilitation.

Algorithms: It was developed to train datasets for statistical applications to ensure data processing with accuracy. These principles form the basis of machine learning (ML), which enables computers to make successful predictions using past experience. While both artificial intelligence and machine learning can enable these advances, such technologies also create safety issues that can cause serious problems for both patients and patients. may raise concerns. all other health care stakeholders. Data privacy and security is one such concern, as most AI applications rely on large amounts of data to make better decisions. Also, ML systems often use data to learn and improve themselves. This makes them more at risk for serious problems like identity theft and data breaches. AI can also be associated with low prediction accuracy, which raises security concerns.

In health services; security means reducing or minimizing the uncertainty of risks and harmful events. With the adoption of artificial intelligence in healthcare, the dimensions of security are changing. With both expected and unexpected harm unlikely, AI and ML have been applied to strengthen security. Risk minimization is therefore key to AI-based applications.

Artificial intelligence plays an important role in increasing knowledge and improving outcomes in healthcare. AI has widespread applications for the prediction and diagnosis of diseases, the processing of large volumes of data and the synthesis of insights, and for maximizing efficiency and outcomes in the medical management of disease states.

There are many benefits to using artificial intelligence in healthcare. AI can be of great help in routine clinical practice and research. Quick and easy access to information, increased accessibility, and reduced errors in diagnosing and treating diseases are key benefits of AI. Predictive diagnostics, precision medicine, and the delivery of targeted therapies are some key areas where AI has made significant advances. Virtual follow-up and consultations are cost- and time-effective. For example, AI-based telemedicine applications provide quality of care and reduce waiting times and the possibility of infection during hospital visits. This ultimately results in high patient satisfaction during treatment.

AI has various applications in diagnostics and decision support. AI gives decision-makers access to accurate and up-to-date information to help them make better decisions in real time. AI also finds application in patient triage. Wearable devices are designed to provide remote monitoring and analysis of vital signs and consciousness index. Algorithms are trained to classify disease conditions according to their severity. Models have been developed to triage patients and predict survival in the prehospital setting. Electronic triage (e-triage) has uses in emergency departments.

However, despite such advantages, there are several limitations to successful adoption and smooth implementation of AI. Evidence-based studies on the efficacy and safety of AI in healthcare are scarce. For example, clinicians often show resistance and reluctance to adopt AI in medical practice. There are also privacy, anonymity, ethical, and forensic concerns regarding the adoption of AI-powered systems in medical practice and research.

Like an individual's right to privacy, medical ethics can be threatened by AI and big data features due to the collection and storage of data from a variety of sources. In addition, the abuse of forensic algorithms by hackers to develop autonomous techniques can put the safety and security of vital information at risk. To avoid such problems, AI research must comply with norms and ethical values.

Any direct or indirect effect of AI on patients or physicians should be minimized using preventive and precautionary safeguards.

13.6 Approaches to Achieving Safety in AI

Causation of predictive models, accuracy of prediction, human effort to label out-of-sample cases, and reinforcement and learning of systems all contribute to making applications safe for use in healthcare.

In the field of health services, efforts should be made to maximize the benefits of artificial intelligence. Experts; it proposes four critical points in this regard: quantifying benefits to enable measurement, building trust for AI adoption, building and developing technical skills, and organizing a system of governance. Data protection legislation should be formulated and strengthened for the collection and processing of data in clinical research.

To add value, accuracy, efficiency, and satisfaction to AI in healthcare, quality standards for AI applications must be clearly defined.

To ensure the safe and effective development and adoption of AI and ML in healthcare; methods, guidelines, and protocols must be formulated. Trust and education; it will enable the full functional integration of artificial intelligence into research and practice in healthcare.

13.7 General Situation in Türkiye

The first academic studies on artificial intelligence in Türkiye date back to the late 1990s. For the first time in Türkiye, it was the drug in which artificial intelligence applications and methods were used in the diagnosis of some diseases (Güvenir et al. 1998) or in cell research in an academic study. One of the first and most interesting case studies on the relationship between logistics and decision-making using fuzzy logic, a method of artificial intelligence simulating human reasoning was reviewed by Cengiz et al. (2003).

As Türkiye lacks qualified human capital in AI, Turkish universities have started offering AI courses or AI degrees to meet the needs. Hacettepe University and University of Economics and Technology (TOBB) became the first universities in Türkiye to offer "Artificial Intelligence Engineering" undergraduate degrees as of mid-2019 (Kasap 2019).

In addition, start-ups and private companies are actively involved in Türkiye's artificial intelligence advancement. Banking and financial services seem to be the main industries using AI applications for their business operations. For instance, İşbank, the largest bank in Türkiye, and Koç University are preparing to establish the "Artificial Intelligence Application and Research Center" to contribute to Türkiye's academic and scientific activities (Garip 2020).

In the field of health; digital transformation has been adopted in management and clinical processes. Every day, new steps are taken in automation. Artificial intelligence applications have adapted very quickly to the health sector. It has different applications in both managerial and clinical processes. AI reduces both administrative and clinical costs by reengineering healthcare processes. It accelerates processes such as diagnosis and treatment in clinical processes and aims to increase service quality by reducing human interaction.

AI applications are heavily used in healthcare applications, from diagnosis of diseases to decision support systems. The Turkish Ministry of Health is actively working to integrate AI applications into Türkiye's healthcare systems. First AI institution, Türkiye Health Data Research and Artificial Intelligence Applications Institute (TUYZE), was established in 2019 under Health Institutes of Türkiye (TÜSEB) to regulate health-oriented AI applications in the state mechanism in Türkiye. The purpose of the Institute is to develop advanced methods, technologies, and value-added products for solving health problems and increasing the efficiency of health services by conducting innovative research activities within the framework of health data research and artificial intelligence applications, to provide courses to train competent human resources in the field of data science and artificial intelligence, to carry out domestic and international cooperation and to carry out the necessary research, training, organization and coordination studies for the creation of our country's digital health ecosystem (TÜSEB 2022).

In addition, while The Scientifc And Technological Research Council Of Türkiye (TÜBİTAK) accepts artificial intelligence as a "prioirity area" in the field of health, it has been giving grants to private companies and public institutions with various scientific and technological projects since 2015. The AI calls for projects reflect both needs as well as expectations in the field of healthcare. Humanoid robots for the treatment process, personalized treatment and AI-oriented medical ventilators were some of the main themes of these calls (TÜBİTAK 2015).

Image analysis used for diagnostic and clinical decision support systems; It is the main element of general and positive expectations regarding artificial intelligence applications of health services in Türkiye. Basically, the expectations focus on preventing human-induced errors and ensuring efficiency in the health system. Türkiye actively uses e-health applications and electronic health records of patients. In this context, national expectations are derived from the effective use of big data and the benefits it provides to the country's economy (Engür et al. 2020).

Digital technologies can exist in various fields such as portability, dressing, communication between machines, cloud computing, internet of things (IoT) and artificial intelligence. Utilizing or using these technologies in health services ensures the digitalization of processes (Altuntaş 2019).

Nowadays, using of artificial intelligence, one of the digital solutions mentioned above, is gaining importance in health services. By using methods such as machine learning and deep learning, which are sub-branches of artificial intelligence, healthcare professionals are moving to new methods in processes such as diagnosis, diagnosis, treatment, rehabilitation and protection of health. These methods provide convenience to health institutions in terms of both cost and health professional competence.

13.7.1 EMRAM (Electronic Medical Record Adoption Model)

It is the criterion model used to determine the level of digital maturity in hospitals. It evaluates hospitals on a scale of 0–7. There are 4 EMRAM level 7 and 62 EMRAM level 6 hospitals in Türkiye (Dijital Hastane 2022).

Level 6 technology is used to implement closed-loop process, drug, blood products and breast milk management, blood sample collection and monitoring. Closed-loop processes are fully implemented in 50% of hospitals. Electronic drug management record (eMAR) and technology are integrated with electronic ordering system (CPOE), pharmacy and laboratory systems to maximize secure point-of-care processes and outcomes.

Level 7 is the hospital to provide and manage patient care; it no longer uses paper and all patient data, medical images and other documents are included in the patient record (EMR) environment. The data pool is used to analyze patterns of clinical data to improve healthcare quality, patient safety and efficiency. Clinical information can be easily shared with all units authorized to treat patients (other non-associated hospitals, outpatient clinics, subacute setting, employer, debtor and patients in the data sharing domain) via standardized electronic transactions (CCD) or a health information exchange. The hospital shows summary data continuity for all hospital services (e.g. inpatient, outpatient, emergency medical documents). Physician documentation and electronic orders are used in 90% (emergency departments not included in this percentile) and closed-loop processes in 95% (emergency departments not included in this percentile).

13.7.2 E-Nabız (Personal Health System)

e-Nabız is an application where healthcare professionals and individuals can access health data (such as analysis reports, examination information, previously written prescriptions) held in different data centers of the country. Individuals can access this application via computer or mobile phone. Through the application, doctors can access the data of the treatments and examinations performed by the patients (with the patient's permission) in different places. This minimizes the time spent in health institutions, prevents unnecessary test requests, reduces costs and at the same time strengthens the doctor-patient relationship (e-Nabız 2022).

13.7.3 Family Physician Information System (AHBS)

With this application, the primary health care services applied to the individual affiliated with the family physician are recorded in a central environment and facilitate their follow-up. It is mandatory to use the application by family physicians. It plays an auxiliary role in the second step and next step decisions about the patient. With this feature, it constitutes a decision support system feature.

13.7.4 Decision Support System (KDS)

By definition, they are applications that help to make the right decision by examining various and models together. Today, KDSs are used in administrative decisions in an integrated manner with the systems that produce data under the Ministry of Health (MoH). The KDS used by the MoH works integrated with the "e-Sağlık" application and can present reports at different levels (E-Sağlık 2022).

13.7.5 Central Physician Appointment System (MHRS)

This application was created in order to enable individuals to reach all hospitals throughout the country for appointment purposes and to use the capacity of Oral and Dental Health Centers and Family Health Centers effectively. The application can be accessed via the internet, mobile application or by calling Alo 182. The main purpose of the application is to reduce the crowd in hospitals and to gather the scattered appointment systems under a single roof. The application is integrated with e-Nabız, e-Devlet applications (MHRS 2022).

13.7.6 Medula

It is a project carried out by the Social Security Institution (SGK). This system was established in order to allocate health services from the state through SGK in private health institutions. In accordance with the Law No. 5510, it is mandatory to be used in all health institutions (Kördeve 2017). It is used in transactions such as invoicing, referral, and reports in all hospitals.

13.8 Conclusion

Alan Turing, however, was one of the first and prominent figures of artificial intelligence as an independent scientific discipline. As a result of his lectures at London Mathematical Society, he wrote a substantial article named "Computing Machinery and Intelligence" in 1950 based upon the idea of whether machines or robots can think, learn and act just like human beings (Turing 2012). The first academic works about AI in Türkiye dates back to the late 1990s. It was the medicine that AI applications and methods were used in an academic study in Türkiye for the first time to diagnose some diseases. Start-ups and private companies, furthermore, have been actively engaging in the AI progress of Türkiye. AI applications have been intensively using in healthcare applications from diagnosing diseases to decision support systems. Ministry of Health of Türkiye is actively working to integrate AI applications into healthcare systems of Türkiye. First AI institution, Turkish Health Data Research and Artificial Intelligence Applications Institute (TUYZE), was established in 2019 under TÜSEB to regulate health-oriented AI applications in the state mechanism in Türkiye.

Electronic Medical Record Adoption Model is the criterion model used to determine the level of digital maturity in hospitals. It evaluates hospitals on a scale of 0–7. There are 4 EMRAM level 7 and 62 EMRAM level 6 hospitals in Türkiye. Level 7 is the hospital to provide and manage patient care; it no longer uses paper and all patient data, medical images and other documents are included in the patient record

(EMR) environment. There are some other digital systems that can support AI-based systems in the future like, E-Nabız (Personal Health System), Family Physician Information System, Decision Support System, Central Physician Appointment System, and MEDULA.

The machines that emerged with the effect of Industry 4.0 in the field of digital health have made our lives easier. With 5.0, it has started to be used on behalf of humanity and for the benefit of humanity with smart devices using AI. Especially after the COVİD-19 pandemic experience, Türkiye, like other countries, has experienced the advantages of digitalization. The need for artificial intelligence-supported applications is increasing rapidly, especially with the aging of the population, the importance given to health and the increase in expectations for better quality health service provision. Beyond that, making predictions by interpreting big data instantly in areas where fast decision support systems are needed, such as pandemics, shows that artificial intelligence is not a luxury but an indispensable technology. Türkiye has significant advantages for artificial intelligence applications, thanks to its existing healthcare system and investments in its digital infrastructure. However, it should also seize the opportunity to become one of the world's leading countries in the field of artificial intelligence by rapidly putting this advantage into practice.

References

Aitken M, Clancy B, Nass D (2017) The growing value of digital health: evidence and impact on human health and the healthcare system. https://www.iqvia.com/insights/the-iqvia-institute/reports/the-growing-value-of-digital-health

Altuntaş EY (2019) Sağlık Hizmetleri Uygulamalarında Dijital Dönüşüm. Eğitim Yayınevi

Atasoy H, Greenwood BN, McCullough JS (2019) The digitization of patient care: a review of the effects of electronic health records on health care quality and utilization. Annu Rev Public Health 40:487–500

Baras JD, Baker LC (2009) Magnetic resonance imaging and low back pain care for medicare patients. Health Affairs (millwood) 28(6):w1133–w1140

Becker's Healthcare (2018) AI with an ROI: why revenue cycle automation may be the most practical use of AI. https://www.beckershospitalreview.com/artificial-intelligence/ai-with-an-roi-why-revenuecycle-automation-may-be-the-most-practical-use-of-ai.html

Buchanan BG (2005) A (very) brief history of artificial intelligence. AI Mag 26(4):53–60

Case N (2018) How to become a centaur. MIT Press. https://doi.org/10.21428/61b2215c

Cengiz K, Ufuk C, Ziya U (2003) Multi-criteria supplier selection using fuzzy AHP. Logist Inf Manag 16(6):382–394. https://doi.org/10.1108/09576050310503367

Char DS, Shah NH, Magnus D (2018) Implementing machine learning in health care—addressing ethical challenges. N Engl J Med 378:981–983

Dijital Hastane (2022) https://dijitalhastane.saglik.gov.tr/

E-Nabız (2022) E-Nabız Hakkında. https://enabiz.gov.tr/Yardim/Index (Erişim Tarihi: 13.10.2020)

Engür D, Basok B, Orbatu D, Pakdemirli A (2020) Uluslararası Sağlıkta Yapay Zeka Kongresi 2020. Kongre Raporu. https://www.researchgate.net/publication/339975029_Uluslararasi_Saglikta_Yapay_Zeka_Kongresi_2020_Kongre_Raporu

E-Sağlık (2022) KSD (Karar Destek Sistemi). https://e-saglik.gov.tr/TR,7079/kds.html

European Commission (2018) 2018 reform of EU data protection rules. https://www.tc260.org.cn/upload/2019-02-01/1549013548750042566.pdf

Francis R, Estlin T, Doran G, Johnstone S, Gaines D, Verma V, Burl M, Frydenvang J, Montano S, Wiens RC, Schaffer S, Gasnault O, DeFlores L, Blaney D, Bornstein B (2017) AEGIS autonomous targeting for ChemCam on mars science laboratory: deployment and results of initial science team use. Sci Robot 2(7). https://doi.org/10.1126/scirobotics.aan4582

Garip E (2020) Is Bank and Koc University establish Artificial Intelligence Application Centre [İş Bankası ve Koç Üniversitesi'nden Yapay Zeka Uygulama Merkezi]. Anadolu Ajansı. https://www.aa.com.tr/tr/sirkethaberleri/finans/is-bankasi-ve-kocuniversitesinden-yapay-zeka-uygulama-ve-arastirma-merkezi/655834

Griffiths S (2016) This AI software can tell if you're at risk from cancer before symptoms appear. http://www.wired.co.uk/article/cancer-risk-ai-mammograms

Güvenir HA, Demiröz G, Ilter N (1998) Learning differential diagnosis of erythemato-squamous diseases using voting feature intervals. Artif Intell Med. https://www.sciencedirect.com/science/article/pii/S0933365798000281

Haenlein M, Kaplan A (2019) A brief history of artificial intelligence: on the past, present, and future of artificial intelligence. Calif Manage Rev 61(4):5–14. https://doi.org/10.1177/0008125619864925

Haraway D (2000) A cyborg manifesto. In: Bell D, Kennedy B (eds) The cyber cultures reader. Routledge, London and New York

Hutson M (2017) This computer program can beat humans at Go—with no human instruction. Science Magazine. https://www.sciencemag.org/news/2017/10/computer-program-can-beat-humans-gono-human-instruction

Jamieson T, Goldfarb A (2019) Clinical considerations when applying machine learning to decision support tasks versus automation. BMJ Qual Saf 2019(28):778–781

Jiang F (2017) Artificial intelligence in healthcare: past, present and future. Stroke Vasc Neurol 2:230–243

Johnson KW, Torres Soto J, Glicksberg BS, Shameer K, Miotto R, Ali M, Dudley JT et al (2018) Artificial intelligence in cardiology. J Am Coll Cardiol 71(23):2668–2679

Kasap S (2019) Syllabus of the first artificial intelligence engineer department in Turkey is ready [Türkiye'nin ilk "yapay zeka mühendisliği" bölümünün müfredatı hazır]. Anadolu Ajansı. https://www.aa.com.tr/tr/egitim/turkiyenin-ilk-yapayzeka-muhendisligi-bolumunun-mufredati-hazir-/153803

Keller AG (2020) Book review: the book of knowledge of ingenious mechanical devices. Br J Hist Sci 8(1):75–77. http://www.jstor.org/stable/4025822

Kommerskollegium National Board of Trade (2012) How borderless is the cloud? https://www.wto.org/english/tratop_e/serv_e/wkshop_june13_e/how_borderless_cloud_e.pdf

Kördeve M (2017) Sağlık Ödemelerinde Yeni Bir Kavram. Medikal Muhasebe. Ç.Ü. Sosyal Bilimler Enstitüsü Dergisi, pp 1–13

Lee J, Korba C (2017) Social determinants of health: how are hospitals and health systems investing in and addressing social needs? Deloitte Center for Health Solutions. https://www2.deloitte.com/content/dam/Deloitte/us/Documents/life-sciences-health-care/us-lshc-addressing-social-determinants-of-health.pdf

Lye CT, Forman HP, Gao R, Daniel JG, Hsiao AL, Mann MK, deBronkart D, Campos HO, Krumholz HM (2018) Assessment of US hospital compliance with regulations for patients' requests for medical records. JAMA Netw Open 1(6):e183014

McCall B (2020) COVID-19 and artificial intelligence: protecting health-care workers and curbing the spread. Lancet Digit Health. 2020(2):e166–e167

McCarthy J, Minsky ML, Rochester N, Shannon CE (1955) A proposal for the Dartmouth summer research project on artificial intelligence. http://jmc.stanford.edu/articles/dartmouth/dartmouth.pdf

Metz C (2019) A.I. shows promise assisting physicians. The New York Times. https://www.nytimes.com/2019/02/11/health/artificial-intelligence-medical-diagnosis.html. Accessed 12 November 2019

MHRS (2022) Merkezi Hekim Randevu Sistemi. https://www.mhrs.gov.tr/hakkimizda.html

Mordvintsev A, Tyka M, Olah C (2015) deepdream. GitHub. https://github.com/google/deepdream
Noor AK (2017) AI and the future of machine design. Mech Eng 139(10):38–43. OpenAI. 2019. OpenAI. https://openai.com
OECD (2019) Artificial intelligence in society. OECD Publishing, Paris. https://doi.org/10.1787/eedfee77-en
Russell SJ, Norvig P (2010) Artificial intelligence a modern approach. In: Artificial intelligence a modern approach. https://doi.org/10.1017/S0269888900007724
Schulte F, Fry E (2019) Death by 1,000 clicks: where electronic health records went wrong. Kaiser Health News & Fortune (joint collaboration). https://khn.org/news/death-by-a-thousand-click
Shin SY (2019) Current status and future direction of digital health in Korea. Korean J Physiol Pharmacol 23(5):311–315
Stanford Engineering (2019) Carlos bustamante: genomics has a diversity problem. https://engineering.stanford.edu/magazine/article/carlos-bustamante-genomics-has-diversity-problem
Topol EJ (2019) High-performance medicine: the convergence of human and artificial intelligence. Nat Med 25(1):44–56
Tran BX, Vu GT, Ha GH, Vuong QH, Ho MT, Vuong TT, Ho R (2019) Global evolution of research in artificial intelligence in health and medicine: a bibliometric study. J Clin Med 8(3):360
Turing AM (2012) Computing machinery and intelligence. In: Machine intelligence: perspectives on the computational model, vol 49, pp 1–28. https://doi.org/10.1093/mind/lix.236.433
TÜBİTAK (2015) Call for healthcare and biomedical equipments [SağlıkBiyomedikal Ekipmanlar Duyurusu]. https://www.tubitak.gov.tr/sites/default/files/1511-sab-bmed-2015-2.pdf
TÜSEB (Türkiye Sağlık Enstitüleri Başkanlığı) (2022) https://www.tuseb.gov.tr/tuyze/about-tuyze
West SM, Whittaker M, Crawford K (2019) Gender, race and power in AI. AI Now Institute. https://www.ainowinstitute.org/discriminatingsystems.html
Xu N, Wang K (2019) Adopting robot lawyer? The extending artificial intelligence robot lawyer technology acceptance model for legal industry by an exploratory study. J Manag Organ. https://doi.org/10.1017/jmo.2018.81
Yu A (2019) How Netflix uses AI, data science, and machine learning—f rom a product perspective. Medium. https://becominghuman.ai/how-netflix-uses-ai-and-machine-learning-a087614630fe

Didem İncegil She was born in 1982 in Ankara, Turkey. She graduated from *Hacettepe University Faculty of Engineering* in 2004 and completed her M.Sc. in 2007. Currently, she is continuing her Ph.D. in *Business Administration at the Faculty of Economics and Administrative Sciences* at Haci Bayram University. Didem İncegil works as an specialist in International Programs Department at the Turkish Health Quality and Accreditation Institute (TÜSKA) within the body of Health Institutes of Türkiye (TÜSEB).

İbrahim Halil Kayral He is a graduate of Hacettepe University Faculty of Economics (Eng) Department. He completed his MBA in Production Management and Marketing at Gazi University in 2007 and Ph.D. with his work on perceived service quality in health enterprises in the Department of Business Administration at the same university. After graduating, he worked on training and consultancy services in World Bank and KOSGEB supported Youth Entrepreneurship Development Projects and Small-Scale Business Consultancy Support Projects. He was appointed as associate professor at İzmir Bakırçay University at Faculty of Health Sciences, Healthcare Management in 2021. He carried out studies on strategic management in the public sector and later on performance-based payment systems, corporate performance and executive performance in the Strategy Development Department of the Ministry of Health. Within the scope of the IAP signed with ISQua in the Department of Quality and Accreditation in Health, he carried out studies to ensure the coordination between the Ministry of Health and ISQua and to establish the infrastructure for accreditation in health in Türkiye. Kayral, who is Deputy Secretary General of Health Institutes of Türkiye (TÜSEB) and has been working at Türkiye Healthcare Quality and Accreditation Institute (TÜSKA) since 2015, also continuous to associate at İzmir Bakırçay

University, Department of Health Management. Kayral, who is an invited speaker and carries out studies in many different countries with his many books, book chapters, articles and papers published nationally and internationally, is an Honorary Member of the International Accreditation Commission Confederation—CIAC Global Advisory Board and a member of the Higher Education Quality Board (YÖKAK). He gives lectures in the fields of health tourism management, quality and accreditation, patient and employee safety, business and strategic management, organizational behavior and behavioral sciences.

Figen Çizmeci Şenel She was born in 1971 in Denizli, Turkey. She graduated from Ankara University Faculty of Dentistry in 1994 and completed her Ph.D. in 2001 at Oral, Maxillofacial and Maxillofacial Surgery Department of the same university. In 2002, she worked as a research fellow in Department of Oral and Maxillofacial Surgery, Washington Hospital Center, USA. At the same year, she completed "Introduction to the principles and practices of clinical trials" certification program at the National Institute of Health, USA. She was appointed as associate professor at Karadeniz Technical University, Faculty of Dentistry in 2009. She worked as a rotationel attending in Washington Hospital Center, Department of Oral and Maxillofacial Surgery and as a researcher in the National Institute of Health, National Institute of Dental and Craniofacial Research in the United States in 2013. During her research, she also received information security awareness, basic information system security authorization, privacy awareness, document and risk management, health records, work station basics, system administration, fundamental rights and discrimination in employees and patients, ethics training.). She was appointed professor at the Faculty of Dentistry of Karadeniz Technical University in 2017. She has been appointed as the chairman of Turkish Health Care Quality and Accreditation Institute in 2018 and still continues her duty. She was appointed Secretary General of Turkish Health Institutes in 2021. Between 2018 and 2022, she served as a member of The National Higher Education Quality Board and chairman of The Commission for the Recognition and Authorization of External Evaluation and Accreditation Bodies. During the Covid-19 Pandemic period, she served as a member of the Turkish National Pandemic Task Force and still continues her duty. She has more than 100 publications in national and international level and 3 books editorial.

Chapter 14
Managing Artificial Intelligence Algorithmic Discrimination: The Internal Audit Function Role

Lethiwe Nzama-Sithole

Abstract Artificial intelligence (AI) systems bring exciting opportunities for organizations to speed up their processes and have a competitive advantage. However, some weaknesses come with some of the AI systems. For example, artificial intelligence bias may occur due to AI algorithms. The algorithms' discrimination or bias may result in organizational reputational risk. This chapter aims to conduct a literature review to synthesize the role of the internal audit function (IAF) in data governance. The chapter will investigate the measures that may be put in place by the IAF to assist the organizations in being socially responsible and managing risks when implementing artificial intelligence algorithms. A literature review will be undertaken using articles recently published with similar keywords for the chapter and the most cited articles from high-impact factor journals. The findings and contributions of the chapter will be updated after the chapter.

Keywords Artificial intelligence · Algorithms discrimination · Internal audit function · AI governance

14.1 Introduction

Artificial intelligence (AI) may provide innovation and productivity in organizations, but it also creates problems. AI-related problems may be due to risk relating to bias, discrimination (Mäntymäki et al. 2022; Peters 2022). Nevertheless, through AI, organizations' operations become more efficient in different areas, such as games, categorization, recommendation, and medical predictions (Tan et al. 2020; Chou et al. 2022). With AI tools, organizations may transform their operations, from setting

L. Nzama-Sithole (✉)
Department of Commercial Accounting, School of Accounting, College of Business and Economics, University of Johannesburg, Johannesburg, South Africa
e-mail: lethiwen@uj.ac.za

M. Kılıç and S. Bozkuş Kahyaoğlu (eds.), *Algorithmic Discrimination and Ethical Perspective of Artificial Intelligence*, Accounting, Finance, Sustainability, Governance & Fraud: Theory and Application, https://doi.org/10.1007/978-981-99-6327-0_14

product prices to extending credit based on customer behavior (Applegate and Koenig 2019).

While AI can broadly be beneficial to many organizations, recent cases have been brought against certain organizations due to discrimination by those affected by it. For example, the U.S. Department of Housing and Urban Development (HUD) alleged that Facebook advertising systems limited their adverts to specific populations based on race, gender, and other characteristics (Allan 2019). The HUD allegedly reported discrimination and indicated that the Facebook systems enabled this discrimination (Allan 2019). Another example is that of Amazon in 2014, where they detected algorithm bias, which resulted in gender discrimination against job applicants (Belenguer 2022). Another case was the Correctional Offender Management Profiling for Alternative Sanction (COMPAS). This COMPAS system was designed to provide the US courts with defendants' risk scores and the likelihood that these defendants could become re-offenders (Belenguer 2022). It was later reported that there was some form of bias when COMPAS was applied, resulting in the wrongful imprisonment of those waiting for trial (Ugwudike 2021; Belenguer 2022). The application of COMPAS also discriminated against black individuals, as they were more likely to be classified as high-risk individuals (Ugwudike 2021; Belenguer 2022). Lastly, bias was also reported in the US healthcare sector, where white patients are given better healthcare than black patients (Tsamados et al. 2021).

Different forms of algorithmic discrimination are reported in the literature, and the most popular ones are biases in the form of race and gender (Allan 2019; Chou et al. 2022). Meanwhile in commercial organizations, both race and gender discrimination are mostly reported on, and this type of discrimination is enabled through facial analysis algorithms (Buolamwini and Gebru 2018; Chou et al. 2022). Race biases are primarily noted in the healthcare sector, where predictive algorithms are widely used (Chou et al. 2022). It is reported that the predictive algorithms often prevent the minority population groups from receiving extraordinary medical care (Chou et al. 2022).

The challenge is that users of systems built on algorithms cannot independently verify the validity and accuracy of data from these AI systems (Buijsman and Veluwenkamp 2022). Therefore, it results in an issue when there could be algorithmic discrimination in the AI systems used by organizations. As such, regulators are broadly concerned about the protection of customers (de Marcellis-Warin et al. 2022). Consequently, it is emphasized that when organizations embrace the opportunities offered by AI algorithms, they need to be vigilant of potential risks of discriminatory algorithms and take action to mitigate these potential risks as it affects people's lives negatively (Allan 2019).

The U.S. Senate bill, the Algorithmic Accountability Act of 2019, would direct the U.S. Federal Trade Commission (FTC) to require large companies to audit their AI Algorithms for bias and correct these (Applegate and Koenig 2019). For such a bill to be passed, it is evident that the risk of algorithmic discrimination is a significant concern that needs to be paid attention to by all organizations around the globe. Thus, organizations should ensure that their systems are free from bias and discrimination and that fair treatment is almost guaranteed for their customers (Koshiyama et al.

2022). In Europe, the European General Data Protection Regulation (GDPR) was passed to prescribe that there should be an audit and verification of decisions from AI systems (Chou et al. 2022). The European Artificial Intelligence Act (AIA) was also proposed as a point of reference for how AI systems may be regulated (Mökander et al. 2021). Mökander et al. (2021) posit that the European AIA may be referred to as the EU approach to AI governance. In the UK, the UK Information Commissioner's Office has guided how AI systems should be audited to address the AI risk (Mökander and Floridi 2021).

This chapter aims to investigate the measures that may be put in place by the Internal Audit Function to assist the organizations in being socially responsible and managing risks when implementing artificial intelligence algorithms. The chapter is organized as follows: The second section provides the chapter's objective and research methodology. An overview of the literature review is covered in the third section under pre-specified research questions. The fourth section presents the bibliometrics of the studies related to algorithm discrimination managed with a focus on the role of the internal audit function in ensuring that they play a role in assisting organizations to manage risks that may be brought by algorithm discrimination or bias. The fifth section of the chapter compares the clusters from the bibliometric analysis and make recommendations based on the results of the bibliometric analysis. Finally, the sixth section presents the main conclusions of the chapter.

14.2 Objective and Research Methodology

For this study to be more focused and identify research gaps in line with the research purpose of identifying the role of the internal audit function in assisting an organization in managing the Artificial Intelligence Algorithms discrimination risks, the following questions are proposed:

- RQ1: What are the main algorithms in organizations that may result in discrimination?
- RQ2: What bias and risks may be brought on by algorithmic discrimination?
- RQ3: What governance measures may be put in place by organizations to mitigate algorithmic discrimination?
- RQ4: What is the role of the internal audit function in managing biases that may be brought on by algorithm discrimination?
- RQ5: What is the bibliometric analysis of studies relating to algorithms in the context of discrimination, bias, audit, and governance?

The methodology utilized in this chapter follows a systemic review analysis approach to investigate algorithm discrimination. This methodology is further utilized in identifying the role of the internal audit function in assisting organizations in managing algorithm discrimination. The impact of algorithm discrimination on data bias is investigated and discussed. The study conducted a literature review and a survey on algorithms in the field of accounting, finance, and auditing. The literature review was conducted to identify the risks that may be brought to the organization due to the use of algorithms. The research questions bullet one to four were answered from conducting literature review through Scopus database and selecting article relating to the keywords of the study.

The Scopus Academic Database was utilized to address the research questions. The underlying reason for utilizing Scopus is that it publishes high quality, primary literature. It is also reported that Scopus is one of the academic databases with excellent coverage of artificial intelligence literature and provides API to retrieve the required data with minimum restrictions (Chou et al. 2022). The below search query was applied to retrieve academic papers in artificial intelligence related to algorithm discrimination, data bias, audit or internal audit function, or internal auditor and governance.

(*Algorithms* AND *discrimination or Bias)* AND (*algorithms and audit*) AND (algorithms and *Governance*).

The query above enabled the researcher to extract the bibliometric information such as the publication titles, abstracts, keywords, year, reference list, funding and many more. The steps followed by the researcher for the bibliometric analysis are illustrated in Fig. 14.1, and the results from VOSviewer are presented in Fig. 14.2. VOS Viewer software was used to visualize the results of the bibliometric networks. VOSviewer is a software tool used for constructing and visualizing bibliometric networks. The publication titles, abstracts, and authors' keywords from results in Scopus were filtered using keywords of interest, namely, algorithm, algorithms discrimination, audit, data bias, and governance.

The survey period is limited to 2005–2022, since most literature available relating to algorithms is recent.

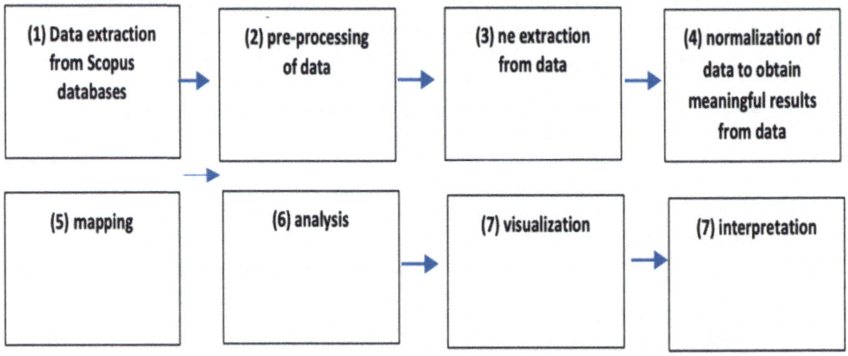

Fig. 14.1 Bibliometric Analysis steps (Martínez et al. 2015; Chen 2017; Kahyaoglu and Aksoy 2021)

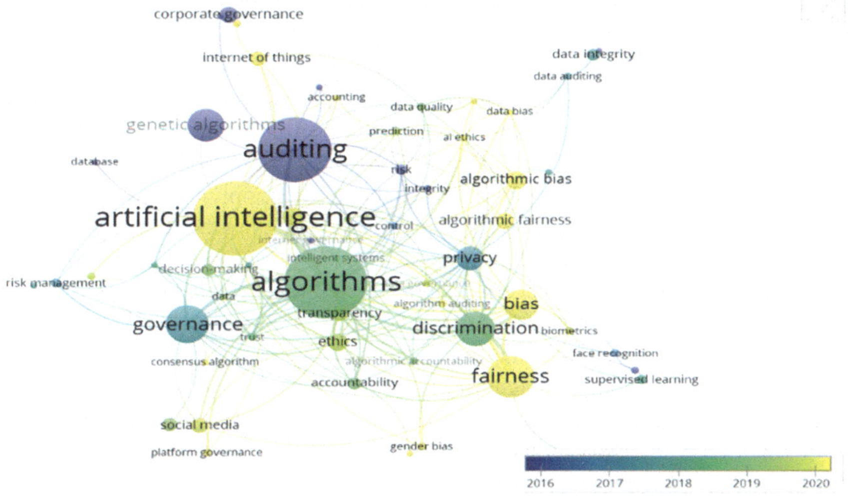

Fig. 14.2 Bibliometric network of Algorithmic discrimination, auditing, and governance. *Source* VOSviewer (2022)

14.3 Related Review of Literature

14.3.1 Algorithms

14.3.1.1 In This Section, the RQ1: What Are the Main Algorithms in Organizations that May Result in Discrimination?

The algorithm is a word that originates from a Persian mathematician whose name was al-Khwarizimi (Belenguer 2022). Belenguer (2022: 3) defines Algorithms as "a set of instructions or rules that will attempt to solve a problem". Pethig and Kroenung (2022) also posit that algorithms can make automated decisions by adapting to and learning from historical data. Furthermore, the learning can take place without requiring programmer intervention (Rodgers and Nguyen 2022).

Haenlein and Kaplan (2019) state that there are four types of AI technologies' namely, genetic algorithms/programming, fuzzy systems, neural networks, and hybrid systems. For the benefit of this chapter, the genetic algorithms will be the focus and will be explained further. Literature indicates two classes of AI algorithms: static and dynamic (Koshiyama et al. 2022). The static algorithms are traditional programs that perform fixed sequences of actions, while the dynamic algorithms progress and embody machine learning (Koshiyama et al. 2022). As a result, algorithms predict the future, which may be based on past experiences or actions.

Sophisticated algorithms may collect a massive amount of data from multiple sources, such as personal information, facial recognition, buying habits, location data, public records, internet browsing habits, and all other information that may be found on electronic devices (Pelletier 2019). These algorithms may result in significant data that may be used for marketing purposes or sold to other organizations without the knowledge of the data owners. Thus, there might be data integrity risks. Pelletier (2019) suggests that there should be controls put in place to ensure that ethics are considered when collecting data and further proposes that internal auditors should play a role in assisting organizations in being ethical when applying algorithms since it suggests that "Algorithms are not ethically neutral" (Tsamados et al. 2021).

The algorithm's predictions may sometimes have some form of discrimination and bias. There have been concerns about the unfairness and bias in AI-based decision-making tools (Landers and Behrend 2022; Belenguer 2022). Belenguer (2022) recommends that when data from the system that uses algorithms are biased, this may lead to discrimination against specific groups or individuals. Many forms of discrimination may occur due to AI bias, including gender, political affiliation, social class, race, or sexual orientation (Belenguer 2022; Peters 2022). Belenguer (2022) further indicates that some form of bias may occur in the organizations, which may need to be dealt with urgently. These include historical bias (this is when bias already exists from the past and is carried forward), representation bias (the sampling of the population and how it is defined, for example, lack of geographical diversity), measurement bias (bias on how we choose, analyze, and measure a particular feature), evaluation bias (incorrect benchmarking), Simpson's Paradox (bias in the analysis of groups and sub-groups), sampling bias, content production bias, and lastly, algorithmic bias.

For the benefit of this chapter, the main aspect focused on is algorithmic biases. The algorithmic bias does not necessarily relate to the data that were put into the system but how the algorithm utilized it (Belenguer 2022).

The algorithmic bias may occur when the unsupervised Machine Learning (ML) models are used. Thus, when raw data are used, the algorithm might find discriminatory patterns, and the system might replicate this, and since there is no human intervention (Rodgers 2020), the discrimination from the system will not be picked up (Belenguer 2022). Belenguer (2022) further argues that when organizations use data mining, there may be discrimination as the application of algorithms decides on their own on which data to value and how to value it.

Algorithmic models expose the organization to substantial risk if there might be possible unethical, illegal, or publicly unacceptable conduct from the decisions made based on the AI models. AI may "damage" customers when algorithms have been used; for example, reduction of customer options and manipulation of customers choices (de Marcellis-Warin et al. 2022).

Belenguer (2022) argues that the quality of data has some form of influence on the quality of algorithmic decisions and suggests that there should be an evaluation of quality and control in place to ensure that algorithms used in the organizations are not biased.

AI is one of the fastest-growing fields globally and is embraced by many organizations (Zemankova 2019; Rodgers and Nguyen 2022). In the past, the field has grown by 270%, and more growth is expected in the future (Zemankova 2019). The growth of AIC is expected to grow even further with the previous estimates; so that by 2023, there may be global spending on AI that may reach an estimated $97.9 billion.

There are a variety of AI definitions available in the literature, with many authors stating their suggested definitions of AI. For example, Haenlein and Kaplan (2019: 5) define AI as "a system's ability to interpret external data correctly, to learn from such data, and to use those learnings to achieve specific goals and tasks through flexible adaptation".

There are also different branches linked to AI: deep learning and machine learning (Haenlein and Kaplan 2019). Machine Learning features algorithms that learn from past patterns and examples to perform tasks (Haenlein and Kaplan 2019; Choong Lee 2019; Koshiyama, et al. 2022). Algorithms and other AI methods assist organizations and individuals in making decisions (Haenlein and Kaplan 2019; Pethig and Kroenung 2022), for example, the decision for prescribing medication, medical treatment, hiring, and buying, which song to listen to, which movie to watch and which friend to connect with (Tsamados et al. 2021).

14.3.2 Risks

14.3.2.1 In This Section, RQ2: What Bias Risks May Be Brought by Algorithm Discrimination?

Risks associated with algorithmic biasness could be damaging to organizations. As highlighted in the introductory section, recent cases have been brought against certain organizations due to discrimination by those affected by AI algorithms. For example, the U.S. Department of Housing and Urban Development (HUD) alleged that Facebook advertising systems limited their adverts to specific populations based on race, gender, and other characteristics (Allan 2019). The HUD allegedly reported discrimination and indicated that the Facebook systems enabled this discrimination (Allan 2019). Another example is that of Amazon in 2014, where they detected algorithm bias, which resulted in gender discrimination against job applicants (Belenguer 2022). As highlighted in the introduction section, the COMPAS system was designed to provide the US courts with defendants' risk scores and the likelihood that these defendants could become re-offenders (Belenguer 2022). It was later reported that there was some form of bias when CAMPAS was applied, resulting in the wrongful imprisonment of those waiting for trial (Ugwudike 2021; Belenguer 2022). The application of COMPAS also discriminated against black individuals, as they were more likely to be classified as high-risk individuals (Ugwudike 2021; Belenguer 2022). As such the COMPUS system brought risks and some form of discrimination to the society and the correctional services. Lastly, bias was also reported in the US healthcare sector, where white patients are given better healthcare than black patients (Tsamados et al. 2021).

Different forms of algorithmic discrimination are reported on in the literature, and the most popular ones are biases in the form of race and gender (Allan 2019; Chou et al. 2022). Meanwhile, in commercial organizations, both race and gender discrimination are mostly reported on, and this type of discrimination is enabled through facial analysis algorithms (Buolamwini 2018; Chou et al. 2022). Race biases are primarily noted in the healthcare sector, where predictive algorithms are widely used (Chou et al. 2022). It is reported that the predictive algorithms often prevent the minority population groups from receiving extraordinary medical care (Chou et al. 2022). These are some of the identified risks that may be addressed through AI governance and the role of Internal Audit Function.

14.3.3 Governance

14.3.3.1 In This Section, RQ3: What Governance Measures May Be Put in Place by Organizations to Mitigate Algorithm Discrimination?

As defined early, AI systems have various benefits for organizations and individuals; some of these benefits are not economical but social (Mökander and Floridi 2022). For example, the use of AI systems in the healthcare sector brings about social benefits for the patient affected, as AI system may assist in solving medical conditions and thus, improve people's lives. However, on the downside, the AI systems may also cause harm to the same patients, when the algorithms of the AI systems have some bias, which may result in discrimination and violation of the same individuals involved in the same system. Therefore, AI governance should be in place to ensure that organizations and individuals do reap the benefits and opportunities that come with the use of these AI systems.

Governance is the combination of processes and structures that assist organizations in achieving their objectives (The Institute of Internal Auditors (IIA) 2018). These governance processes and structures are not only due to the risks that may occur and affect the organizations but may also be due to organizations putting effort into mitigating other potential unknown risks (IIA 2018). Internal auditing assists in ensuring that governance within the organization is in place. This role is crucial, as the internal audit provides objective assurance and insights about the adequacy and effectiveness of risk management, internal control, and governance processes (IIA 2018). When the internal audit function is vibrant and agile, it will be of value to the organization's governance (IIA 2018).

Mäntymäki et al. (2022) post that AI governance needs to be in place to enable the organization to reap the rewards and opportunities, as well as manage the risks of the AI system. It is further suggested that there should be proper governance in place to mitigate any risks associated with AI Governance. As such, AI Governance might result in result in the alignment of the organization in conjunction with human and societal values (Mäntymäki et al. 2022). For good governance to be in place in organizations that adopt AI, regulations and transparency should exist concurrently (Mökander and Floridi 2021). Mökander and Floridi (2021) further argue that there will be the possibility of identifying risks earlier or before they occur and preventing harm to those involved when AI governance is in place.

The AI governance shares a similar characteristic with the governance definition stated above. Butcher and Beridze (2019), cited in Mäntymäki et al. (2022; p. 123), define AI governance as "a variety of tools, solutions and levels that influence AI development and applications" and this definition also links directly to the definition by Gahnberg (2021, p. 123 who defines AI governance as "an intersubjectively recognised rules that define, constrain and shape expectations about the fundamental properties of an artificial agent". While Schneider et al. (2020) in Mäntymäki et al. (2022, p. 123) define AI governance as "the structure of riles, practices, processes used to ensure that the organisation's AI technology sustains and extends the organisational strategies and objectives".

From the definitions of AI governance provided above, there are similar concepts that come up, namely rules, tools, and processes. These should be in place for the proper use/application of AI systems.

Mäntymäki et al. (2022) suggest that an AI governance definition should be action-oriented and be able to guide the organization on how to implement AI systems effectively. Mäntymäki et al. (2022) further suggest that AI has three interlinked areas that need to be considered: corporate governance, IT, and data governance. AI governance overlaps with data governance, both in the center of IT governance (Mäntymäki et al. 2022).

14.3.4 Internal Audit Function

14.3.4.1 In This Section, RQ4: What is the Role of the Internal Audit Function in Managing Biases that May Be Brought by Algorithm Discrimination?

It is recommended for organizations to control factors that may expose them to discriminatory risks, when their operations depend on artificial intelligence models (Allan 2019). Using algorithmic tools in the organizations may bring about risks or potential risks that need to be mitigated in the early stages or even before their occurrence. Belenguer (2022) suggests that when organizations plan to remove biases in their algorithmic models, they need to firstly identify reasons for the occurrence of these biases. Thus, the question may be who should be identifying these possible biases.

Allan (2019) argues that it cannot be the responsibility of the algorithm model developer, as internal control principles dictate that the person who creates a system cannot be the impartial evaluator of the same system. It is then debatable who in the organization should be responsible for managing the possible risks that the AI algorithm's discrimination may bring. Allan (2019) further suggests that internal auditors may be best suited to assure the compliance of the AI algorithm system. Similar views have been shared by Raji et al. (2020), Vanian (2021), and Landers and Behrend (2022), who also propose that auditing may play a significant role in verifying that the AI-driven predictions are fair, unbiased, and valid. De Marcellis-Warin et al. (2022) suggest that auditing may be used to help organizations verify assertions made by AI system developers and those who use the system. Similar views are also shared by Raji and Buolamwini (2019), who posit that the IAF may assist in checking the engineering process involved in developing the AI algorithm system. Furthermore, Brundage et al. (2020) also argue that external auditors might confirm assertions made by AI system developers.

Literature does not explicitly indicate the type of audit or auditor that should be responsible for verifying compliance of AI algorithms used in an organization. The audit will be limited to the internal audit perspective and the internal audit function. De Marcellis-Warin et al. (2022) argue that the audit of AI system should be in place and that the different types of audits are not mutually exclusive, but rather complement each other. However, for the benefit of the focus of this chapter, the audit will be representative of the internal auditor or the internal audit function.

The internal auditors need to ensure whether the predicated outcomes by algorithms used are reasonable (Applegate and Koenig 2019). Thus, internal auditors are needed when there is a potential risk (Seago 2018). The same views are shared by Pelletier (2019), who argues that the risk of unfair biases caused by algorithms may be scrutinized by internal auditors to ensure fair treatment and transparency for the customers of businesses. Allan (2019) suggests that internal auditors may help the organization's management by providing risk-based, objective assurance, advice, and insight. Internal auditors promote trust and transparency (LaBrie and Steinke 2019; de Marcellis-Warin et al. 2022).

In providing the assurance role relating to AI models, the auditors should learn and adapt their methods to meet organizational challenges by adopting AI. When auditors verify the AI algorithms and pick up issues, they may recommend corrective actions, which will add value to the organization (Landers and Behrend 2022). Thus, this complies with one of the principles of Professional Due Care and the Competency Code of Ethics.

The audit of AI system is not only limited to technical performance but also ethical compliance (Sandvig et al. 2014; Diakopoulos 2015; Mökander and Floridi 2021; Ugwudike 2021). Auditors are needed in an organization to assure management and the audit committee that the AI models chosen, do not discriminate (Allan 2019). Allan (2019) further suggests that internal auditors may assist the organization in mitigating reputational, financial, and legal risks caused by implementing a system of algorithmic discrimination or bias. Internal auditors will also assist organizations by providing ethical conscience for their leaders and enhancing their professional responsibility (Applegate and Koenig 2019).

Internal auditors provide insight, as they act as catalysts for the management of the organizations, and they may also provide foresight to the organization by identifying trends. For example, when the internal audit function is mature, it may be proactive and be able to predict future potential risks or challenges that may be faced by the organization (IIA 2018). As such an internal auditor assures a variety of stakeholders, who might have different roles in an organization, for example, regulators (to verify whether algorithms used by the business are legal and meet required standards) and customers, investors, and systems providers for them to make informed decisions of associating themselves with the organizations.

Applegate and Koenig (2019) suggest that an audit of the integrity of data sets should be conducted to ensure that the discrimination risks are mitigated. When data sets are not representative of the actual population, there is a risk of bias (Applegate and Koenig 2019). Applegate and Koenig (2019) suggest that there should be proper controls put in place by the organizations, and they recommend that there should also be an audit test of controls conducted by auditors.

Applegate and Koenig (2019) suggest that internal auditors also need a framework to review the implementation of AI models. Landers and Behrend (2022) theorize that when high-quality auditing processes are applied for the evaluations of the algorithms, the auditor does not only assist in the improvement of algorithmic success but also assists in confirming the claims made by systems developers and assists organizations in improving their public trust and transparency with the organization's use of AI algorithms. Belenguer (2022) similarly suggests that when an assessment of bias is conducted, it will assist organizations in understanding complex social and historical contexts better and further recommends this assessment will assist organizations in improving their ethical framework. Professional accounting firms (Big 4) developed assurance procedures that organizations may perform to ensure that the AI systems they design are safe, legal, and ethical, and these audit procedures may be adopted by internal audit functions when providing their services.

14.3.5 Bibliometric Analysis

14.3.5.1 In This Section, RQ5: What Is the Bibliometric Analysis of Studies Relating to Algorithms in the Context of Discrimination, Bias, Audit, and Governance?

The bibliometric results are summarized based on "co-occurrence" aspect for 2005–2022 period of all articles from Scopus database. A total of five basic clusters emerged. The analysis has 59 items, 5 clusters, 229 links, and 419 strength of the "co-occurrence" mapping. The keywords of the clusters are presented in Appendix A and discussion of the results is presented in the following section:

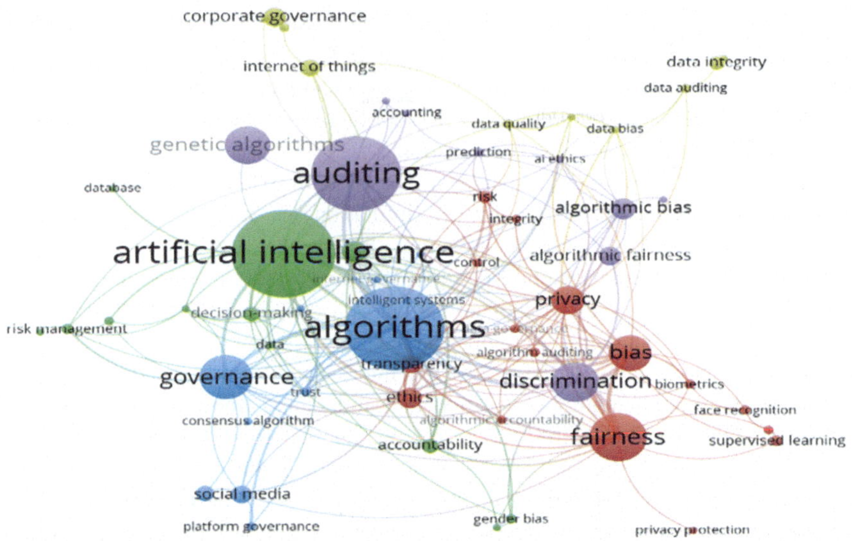

Fig. 14.3 Bibliometric network visualization of clusters per color group. *Source* VOSviewer (2022)

Cluster 1: Fairness (Red)

The first cluster of the bibliometric analysis is presented in Fig. 14.3 and includes keywords such *as "algorithm auditing", algorithmic accountability", "bias", "biometrics", "control", "integrity", "privacy", "face recognition", "risk", unsupervised learning".*

Cluster 2: Artificial Intelligence (Green)

The second cluster of bibliometric analysis is presented in Fig. 14.3 and includes keywords such as *"accountability", algorithmic governance", "compliance", "data", "decision making", "gender bias", risk management"*. From the bibliometric analysis, it may be concluded that when AI is used in an organization, there might be a need to manage risks such as gender biases and organizations would need to comply and algorithmic governance in place.

Cluster 3: Algorithms (Blue)

The third cluster of bibliometric analysis is presented in Fig. 14.3 and includes keywords such as *"AI governance", "governance", "trust"*.

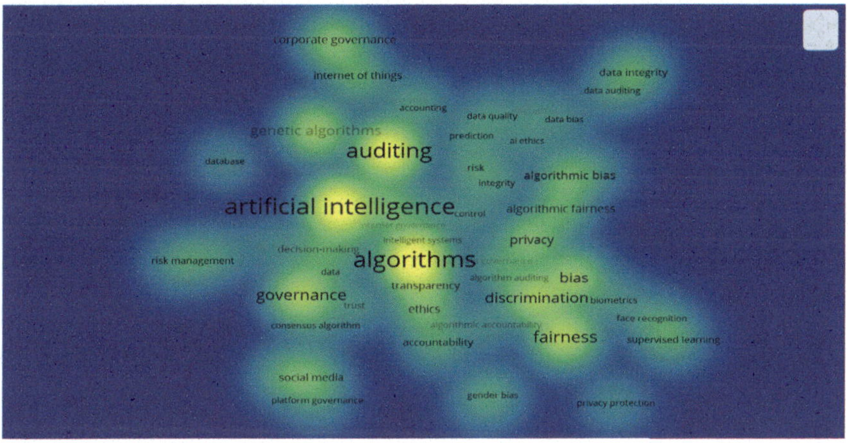

Fig. 14.4 Bibliometric density visualization. *Source* VOSviewer (2022)

Cluster 4: Corporate Governance (Yellow)

The third cluster of bibliometric analysis is presented in Fig. 14.3 and includes keywords such as "corporate governance", "data auditing", "data ethics", "data bias", "data integrity".

Cluster 5: Auditing (Purple)

The third cluster of bibliometric analysis is presented in Fig. 14.3 and includes keywords such as *"accounting", algorithmic bias", "discrimination", "prediction"* (Fig. 14.4).

14.4 Conclusion and Recommendations

It is clear from the research covered in this chapter that, while AI may be critical for organizations, it has its weaknesses that must be closely monitored and addressed. Among these are the well-documented algorithm biases and/or discriminations that AI systems may come along with. In this regard, this chapter conducted a systematic literature review to determine the role the internal audit function can play in assisting organizations in managing AI algorithmic discrimination. This chapter also analyzed the internal audit function's measures that needed to be put in place. In addition, it is evident from the bibliometric analysis that there is growing research interest in linkages in areas of AI, algorithms, governance, and auditing. This chapter could thus give guidance to organizations and relevant practitioners on how internal audit function could be utilized to guard against a myriad of risks associated with AI algorithms.

Appendix

See Table 14.1.

Table 14.1 Bibliometric network analysis—mapping of co-occurrences clusters

Cluster 1 (16 items)	Cluster 2 (12 items)	Cluster 3 (11 items)	Cluster 4 (10 items)	Cluster 5 (10 items)
Algorithm auditing	Accountability	AI governance	Audit logs	Accounting
Algorithmic accountability	Algorithmic governance	Algorithms	Corporate governance	AI ethics
Bias	Artificial intelligence	Consensus algorithm	Data auditing	Algorithmic bias
Biometrics	Compliance	Governance	Data bias	Algorithmic fairness
Control	Data	Intelligent systems	Data ethics	Analytics
Data governance	Data science	Internet governance	Data integrity	Auditing
Ethics	Database	Opinion mining	Data quality	Discrimination
Face recognition	Decision-making	Platform governance	Industry 4.0	Genetic algorithms
Fairness	Explainable AI	Regulation	Internet of things	Prediction
Integrity	Gender bias	Social media	Knowledge management	Social networks
Privacy	Information system	Trust		
Privacy protection	Risk management			
Risk				
Supervised learning				
Supervised learning				
Unsupervised learning				

References

Allan S (2019) Bias in the machine. IA Global, Florida, USA. https://internalauditor.theiia.org/en/articles/2019/june/bias-in-the-mache/

Applegate D, Koenig M (2019) Framing AI audits. Florida, USA. https://internalauditor.theiia.org/en/articles/2019/december/framing-ai-audits/

Belenguer L (2022) AI bias: exploring discriminatory algorithmic decision-making models and the application of possible machine-centric solutions adapted from the pharmaceutical industry. AI Ethics 1–17. https://doi.org/10.1007/s43681-022-00138-8

Brundage M, Avin S, Wang J, Belfield H, Krueger G, Hadfield G, Anderljung M et al (2020) Toward trustworthy AI development: mechanisms for supporting verifiable claims. https://doi.org/10.48550/arXiv.2004.07213

Buijsman S, Veluwenkamp H (2022) Spotting when algorithms are wrong. Minds Mach 1–22. https://www.link.springer.com/article/10.1007/s11023-022-09591-0

Buolamwini GS (2018) Intersectional accuracy disparities in commercial gender classification. Proc Mach Learn Res 81:1

Buolamwini J, Gebru T (2018) Gender shades: intersectional accuracy disparities in commercial gender classification. In: Conference on fairness, accountability and transparency. PMLR, pp 77–91

Butcher J, Beridze I (2019) What is the state of artificial intelligence governance globally? RRUSI J 164(5–6):88–96. https://doi.org/10.1080/03071847.2019.16942-60

Chen C (2017) Science mapping: a systematic review of the literature. J Data Inf Sci 2(2):1–40. https://doi.org/10.1515/jdis-2017-0006

Choong Lee Y (2019) Stronger assurance through machine learning. Florida, United States of America: September. IIA Global. https://internalauditor.theiia.org/en/articles/2019/september/stronger-assurance-through-machine-learning/

Chou YL, Moreira C, Bruza P, Ouyang C, Jorge J (2022) Counterfactuals and causability in explainable artificial intelligence: theory, algorithms, and applications. Inf Fusion 81:59–83

de Marcellis-Warin N, Marty F, Thelisson E, Warin T (2022) Artificial intelligence and consumer manipulations: from consumer's counter algorithms to firm's self-regulation tools. AI Ethics 2(2):259–268. https://doi.org/10.1007/s43681-022-00149-5

Diakopoulos N (2015) Algorithmic accountability: journalistic investigation of computational power structures. Digit Journal 3(3):398–415. https://doi.org/10.1080/21670811.2014.976411

Gahnberg C (2021) What rules? Framing the governance of artificial agency. Policy Soc 40(2):194–210. https://doi.org/10.1080/14494035.2021.19297-29

Haenlein M, Kaplan A (2019) A brief history of artificial intelligence: on the past, present, and future of artificial intelligence. Calif Manag Rev 61(4):5–14. https://doi.org/10.1177/0008125619864925

Kahyaoglu SB, Aksoy T (2021) Survey on blockchain based accounting and finance algorithms using bibliometric approach. In: Alsharari NM (ed) Accounting and finance innovations. https://books.google.co.za/books?hl=en&lr=&id=27ZaEAAAQBAJ&oi=fnd&pg=PA35&dq=Survey+on+Blockchain+Based+Accounting+and+Finance+Algorithms+Using+Bibliometric+Approach&ots=FVATjEwvIl&sig=jf2BlNDGSQImO16mE7AuPAcdvKY&redir_esc=y#v=onepage&q=Survey%20on%20Blockchain%20Based%20Accounting%20and%20Finance%20Algorithms%20Using%20Bibliometric%20Approach&f=false

Koshiyama A, Kazim E, Treleaven O (2022) Algorithm auditing: managing the legal, ethical and t technological risks of artificial intelligence, machine learning and associated algorithms. IEEE Comput Soc. https://doi.org/10.1109/MC.2021.3067225

LaBrie RC, Steinke G (2019) Towards a framework for ethical audits of AI algorithms. https://aisel.aisnet.org/amcis2019/data_science_analytics_for_decision_support/data_science_analytics_for_decision_support/24/

Landers RN, Behrend TS (2022) Auditing the AI auditors: a framework for evaluating fairness and bias in high stakes AI predictive models. Am Psychol. https://doi.org/10.1037/amp0000972

Mäntymäki M, Minkkinen M, Birkstedt T, Viljanen M (2022) Defining organizational AI governance. AI Ethics 1–7. https://doi.org/10.1007/s43681-022-00143-x
Martínez MA, Cobo MJ, Herrera M, Herrera-Viedma E (2015) Analyzing the scientific evolution of social work using science mapping. Res Soc Work Pract 25(2):257–277. https://doi.org/10.1177/1049731514522101
Mökander J, Floridi L (2021) Ethics-based auditing to develop trustworthy AI. Minds Mach 31(2):323–327. https://doi.org/10.1007/s11023-021-09557-8
Mökander J, Floridi L (2022) Operationalising AI governance through ethics-based auditing: an industry case study. AI Ethics 1–18. https://doi.org/10.1007/s43681-022-00171-7
Mökander J, Axente M, Casolari F, Floridi L (2021) Conformity assessments and post-market monitoring: a guide to the role of auditing in the proposed European AI regulation. Minds Mach 32:241–268. https://doi.org/10.1007/s11023-021-09577-4
Pelletier J (2019) Internal audit and data ethics. IIA Global, Florida, United States of America. https://internalauditor.theiia.org/en/voices/blog/pelletier/2019/internal-audit-and-data-ethics/
Peters U (2022) Algorithmic political bias in artificial intelligence systems. Philos Technol 35(2):1–23. https://doi.org/10.1007/s13347-022-00512-8
Pethig F, Kroenung J (2022) Biased humans, (un) biased algorithms? J Bus Ethics 1–16. https://doi.org/10.1007/s10551-022-05071-8
Raji ID, Buolamwini J (2019) Actionable auditing: investigating the impact of publicly naming biased performance results of commercial AI products. In: Proceedings of the 2019 AAAI/ACM conference on AI, ethics, and society, pp 429–435. https://doi.org/10.1145/3306618.3314244
Raji ID, Smart A, White RN, Mitchell M, Gebru T, Hutchinson B, Barnes P et al (2020) Closing the AI accountability gap: defining an end-to-end framework for internal algorithmic auditing. In: Proceedings of the 2020 conference on fairness, accountability, and transparency, pp 33–44. https://doi.org/10.1145/3351095.33728-73
Rodgers W (2020) Artificial intelligence in a throughput model: some major algorithms. Science Publishers (CRC Press). https://doi.org/10.1201/9780429266065
Rodgers W, Nguyen T (2022) Advertising benefits from ethical artificial intelligence algorithmic purchase decision pathways. J Bus Ethics 1–19. https://doi.org/10.1007/s10551-022-05048-7
Sandvig C, Hamilton K, Karahalios K, Langbort C (2014) Auditing algorithms: research methods for detecting discrimination on internet platforms. In: Data and discrimination: converting critical concerns into productive inquiry, vol 22, pp 4349–4357. https://social.cs.uiuc.edu/papers/pdfs/ICA2014-Sandvig.pdf
Schneider J, Abraham R, Meske C (2020) AI governance for businesses. https://doi.org/10.48550/arXiv.2011.10672
Seago J (2018) Behind data. Florinda, United States of America. htttps://www.internalauditor.theiia.org/en/articles/2018/april/behind-the-data/
Tan W, Tiwari P, Pandey HM, Moreira C, Jaiswal AK (2020) Multimodal medical image fusion algorithm in the era of big data. In: Neural computing and applications, pp 1–21. https://www.link.springer.com/article/10.1007/s00521-020-05173-2
The Institute of Internal Auditors (IIA) (2018) Internal auditor's role in corporate governance. IIA Global
Tsamados A, Aggarwal N, Cowls J, MorleyJ, Roberts H, Taddeo M, Floridi L (2021) The ethics of algorithms: key problems and solutions. AI Soc 37:215–230. https://doi.org/10.1007/s00146-021-01154-8
Ugwudike P (2021) AI audits for assessing design logics and building ethical systems: the case of predictive policing algorithms. AI Ethics 1–10. https://doi.org/10.1007/s43681-021-00117-5
Vanian J (2021) Federal watchdog says A.I. vendors need more scrutiny. Fortune. https://fortune.com/2021/07/13/federal-watchdog-a-i-vendorsneed-more-scrutiny/
VOSviewer 1.6.16 program (2022) https://www.vosviewer.com/

Zemankova A (2019) Artificial intelligence in audit and accounting: development, current trends, opportunities, and threats-literature review. In: 2019 international conference on control, artificial intelligence, robotics and optimization (ICCAIRO). IEEE, pp 148–154. https://doi.org/10.1109/ICCAIRO47923.2019.00031

Dr. Lethiwe Nzama-Sithole is a Senior Lecturer in the Department of Commercial Accounting at the University of Johannesburg. Lethiwe teaches Auditing and Internal Controls at a second-year level in the Diploma in Accountancy. She has previously lectured on Financial Accounting, worked as an Accountant and Internal Auditor and has over 18 years of working experience. She also teaches at the Johannesburg Business School on a part-time basis. Lethiwe is a well-balanced academic, as she contributes immensely to the Accountancy sector through teaching and learning, research activities and community service. She is a Certified Internal Auditor and holds a Ph.D. in Auditing. She supervises MBA, M.Com and B.Com Honours students. She also serves as a moderator and examiner for other universities. Her research interests are Governance and Auditing, Computer Auditing, ERP Systems and Accounting Education.

Author Index

M. Kılıç and S. Bozkuş Kahyaoğlu (eds.), *Algorithmic Discrimination and Ethical Perspective of Artificial Intelligence*, Accounting, Finance, Sustainability, Governance & Fraud: Theory and Application, https://doi.org/10.1007/978-981-99-6327-0

Subject Index

M. Kılıç and S. Bozkuş Kahyaoğlu (eds.), *Algorithmic Discrimination and Ethical Perspective of Artificial Intelligence*, Accounting, Finance, Sustainability, Governance & Fraud: Theory and Application, https://doi.org/10.1007/978-981-99-6327-0